D1539114

Sponges

Sponges

Patricia R. Bergquist

University of California
Berkeley and Los Angeles

University of California Press
Berkeley and Los Angeles

ISBN: 0-520-03658-1

Library of Congress Catalog Card Number: 77-93466

Printed in Great Britain

Contents

Introduction

It is easy to define a sponge; it is a sedentary, filter-feeding metazoan which utilizes a single layer of flagellated cells (choanocytes) to pump a unidirectional water current through its body.

It is not so easy, without a good deal of qualification, to make further generalizations that characterize all groups that are at present included in the Porifera. If we exclude the Hexactinellida, or 'glass sponges', where both structure and biology are very little known, then it is possible to make further general statements with certainty.

The sponge body is isolated from the external environment by a perforated epithelium, one cell thick. Both the internal flagellated epithelium (the choanoderm) and the external epithelium (the pinacoderm) differ from other metazoan epithelia in that they lack a stable basement membrane. Between these two thin layers is a third region, the mesohyl, which can vary in composition and extent, but which always includes some mobile cells and some skeletal material.

This basic, simple sponge organization has proved a successful and persistent one in evolution. In recent seas, sponges number about 5000 species and account for much of the epifaunal biomass. In Palaeozoic times, sponges in reef locations exceeded the combined biomass totals of other benthic animals (Finks 1970). With the possible exception of one class, the Calcarea, all major types of sponge organization were established and diversified during Cambrian times, and Calcarea were certainly present in the late Palaeozoic.

Historically, the classes of Porifera have been defined by the chemical nature of the inorganic skeletal material, in conjunction with the structure of the skeletal components. Classification will be outlined in detail in a later chapter, but in introduction the outlines of classification at the class level should be understood.

At the present time most workers recognize four classes of recent Porifera: Calcarea, Hexactinellida, Demospongiae and Sclerospongiae.

The Calcarea are exclusively marine sponges with a skeleton of calcium carbonate, organized either as discrete spicules, or as a fused mass. The crystalline structure of the skeleton is usually calcitic. Spicules in the Calcarea are not differentiated in terms of size into the large megasclere and small microsclere categories found in other classes.

The Hexactinellida are again marine sponges which are more common in deep water, their skeleton is siliceous, made up of megascleres and microscleres, both of which can have a hexactine structure. Such spicules have three main axes and hence six points, or actines. Many reduced forms exist, but it is conjectured that this hexactinal structure is primitive in the class. Frequently the megasclere skeleton is an elaborate silica network over which the sparse soft tissue is dispersed. Hexactinellida are reported to lack a superficial cellular pinacoderm and a mesohyl matrix.

The Demospongiae is the largest group of sponges, accounting for 95 % of recent species. Representatives can be found in all aquatic environments: fresh and brackish waters, intertidal to the deepest ocean trenches, polar to tropical seas. Demospongiae are characterized by possession of a skeleton of one- (monactine) to four- (tetractine) rayed siliceous spicules in conjunction with, or supplementary to, an organic skeleton of collagen which can be dispersed in the mesohyl, or secreted in the form of large fibres or filaments of spongin.

The fourth class, the Sclerospongiae, was established to receive sponges which have siliceous spicules and an organic collagen skeleton both restricted to a thin superficial veneer of living tissue supported upon a massive calcareous skeleton, which can be either aragonitic or calcitic. Siliceous spicule pseudomorphs are formed in this calcareous mass as spicules are enveloped by the spread of the massive skeleton. In the structure of their calcareous skeleton, which we shall consider later, the sclerosponges bear many resemblances to the Palaeozoic reef-building Stromatoporoidea, and also to some tabulate corals. It remains a matter of debate at this time whether these sponges should be held distinct at the class level from Demospongiae, or whether the concept of Demospongiae should be widened to admit forms which have supplementary calcareous skeletons. The latter option would not be unreasonable, since the Demospongiae already include (1) forms which have only a fibrillar, dispersed collagenous skeleton (*Oscarella*); (2) forms with dispersed plus fibrous collagenous skeleton (*Spongia*); (3) forms with dispersed collagen, fibre and spicule skeleton (*Clathria*); and (4) forms, like *Timea* and *Hamacantha*, where a basal calcareous skeleton is sometimes present in addition to the collagen and spicule skeleton. In this book we will consider the Demospongiae and Sclerospongiae as separate classes, an interpretation which primarily recognizes that there are basic structural differences and that the groups have had a long and separate fossil history.

The fact that classification of sponges can still be debated at the class level, makes it apparent that there is still much to learn about these organisms. Indeed, it is reasonable to question whether the Hexactinellida should remain in the same phylum as the other groups – the reported absence of a pinacoderm is a very basic difference, since the presence of this layer, and the extent of its development, are essential to any definition of individuality in sponges. Also, without a cellular pinacoderm there can be no pores in the true sense, likewise a basic attribute for a poriferan. In the course of the long fossil history of the Porifera, there has been time for many structural experiments based on a flagellated-cell-powered filter-feeding organization. Perhaps the Hexactinellida should be considered, like the Archaeocyatha, to be a separate but related evolutionary development. It is difficult to argue one way or the other at present because lack of information precludes evaluation of the relationships of the Hexactinellida.

Levi (1956) in his study of development and biology of some Demospongiae found it necessary to state that the Porifera was the last major group of organisms in which the orders were not clearly defined. Since 1956, the orders of Demospongiae and Calcarea have been defined, but doubts about higher level classification have become more acute.

Why does this state of affairs persist? What is so problematic about sponges?

There are several reasons why our knowledge of the biology, and hence of the relationships between sponges, has grown so slowly, despite great interest in the group by nineteenth-century biologists who were concerned mainly with debating developmental theories, or establishing the animal as opposed to the plant nature of sponges.

First, sponges, with the exception of some fresh-water forms, are difficult to sustain under experimental conditions. They may survive for long periods, but their activity is almost invariably far below normal. Consequently sponges have proved unsatisfactory material for experimental and physiological studies. Despite the obvious significance of sponges as model systems in tissue culture and cell reaggregation, until the 1950s experimental work on whole organisms attracted the attention of only a few patient workers.

Secondly, most early work of a systematic or morphological nature underlined the great plasticity of sponges in response to local and geographic environmental variables. Almost all attributes of sponge structure were considered to vary within wide limits, thus making description and definition of species a specialist task. This problem has been overemphasized, but it remains true that there are more than usual difficulties in defining a sponge species. Even today, there is no objective test which can be applied to sponges in order to decide whether similar, discontinuously distributed forms are different species and reproductively isolated, or are variants of a single species. There is tremendous opportunity within the Porifera for ecological studies relevant to this problem.

Thirdly, because so many features of sponge morphology are suspected of variation which is unrelated to genetic differences, it has made the choice of characteristics to be used in delimiting major systematic categories very difficult, and more subjective than in many other invertebrate groups. The consequent lack of a stable classification has made sponges difficult and unpredictable material for the ecologist and physiologist.

These problems have combined over a long period to ensure that students of sponges were few in number.

With the general expansion of biology over the last twenty-five years sponges have begun to command more attention. Modern techniques, particularly electron microscopy, histochemistry, the use of radioactive tracers and scuba diving have proved more informative and versatile with respect to sponge material than earlier and more conventional methodology. Also, as molecular biology and biochemistry have developed they have made available to marine biologists sensitive, but routine, separation techniques which have made biochemical approaches to taxonomy possible. This holds great promise for the Porifera where, relative to other groups, morphology provides so few characters for description.

Consequently, we have seen a great increase in knowledge of sponges, and, although many quite elementary problems are still unresolved, it is now possible to write a book which deals in a comparative way with the biology of the Porifera.

Sponge organization represents a peculiar, specialist evolutionary strategy and, in this volume, I will take wherever possible a functional approach to describing and interpreting the interactions which take place within a sponge and between a sponge and its environment.

It is beyond the scope of this book to examine any particular aspect of sponge biology in great detail and therefore considerable selection will be exercised in choice of material. As a general rule, I will not review the older literature and trace the development of ideas, but will concentrate on presenting and evaluating modern studies on all aspects of sponge biology.

It has been usual in textbooks of invertebrate zoology to base the treatment of the Porifera on the structure and organization of calcareous sponges that are often very simple. More complex Demospongiae have always been considered very briefly. In the following chapters, I will give primary consideration to the Demospongiae, which are by far the most diverse and ecologically significant group. These are the sponges encountered most frequently on the shore and hence in laboratories. Throughout the text, general statements apply to Demospongiae and where the Calcarea differ markedly they will be mentioned separately. The Hexactinellida in almost all aspects require to be treated separately.

The first chapter deals with sponge organization and function: the essential sponge and the elaborations of structure to be found within the phylum. An examination of the range of form and organization allows a consideration of the old question 'What constitutes a sponge individual?' In relation to cellular structure and diversity, I will consider in Chapter 2 the level of differentiation and the plasticity of sponge cells and the behaviour of these cells in dissociated and aggregating suspensions. An integral component of the sponge body is the skeleton, organic and mineral, and this is considered in Chapter 3. Chapter 4 deals with reproduction and development. With these topics outlined, it is possible in Chapter 5 to detail and interpret the modern classification of sponges. The ecology of adult and larval sponges is dealt with in Chapter 6. Chapter 7 deals with some aspects of sponge biochemistry and Chapter 8 with the evolution of sponges, their fossil history and phylogenetic relationships. Present views on their relationships of sponges to other groups of Metazoa will be considered in a brief, final chapter.

Table A: *Classification of all the recent sponge genera cited in the text with the exception of those which occur only in the systematic section of Chapter 3*

Class	Subclass	Order	Genus
Hexactinellida	Amphidiscophora	Amphidiscosa	*Hyalonema*
			Semperella
	Hexasterophora	Lyssacina	*Euplectella*
			Lophocalyx
			Rossella
		Dictyonina	*Farrea*
Calcarea	Calcinea	Clathrinida	*Clathrina*
			Ascandra
		Leucettida	*Leucettusa*
	Calcaronea	Leucosolenida	*Leucosolenia*
		Sycettida	*Sycon*
			Aphroceras
			Leucandra
	Pharetronida	Inozoa	*Petrobiona*
			Murrayona
			Paramurrayona
Demospongiae	Homoscleromorpha	Homosclerophorida	*Oscarella*
			Plakina
			Plakortis
	Tetractinomorpha	Choristida	*Ancorina*
			Geodia
			Stelletta
			Disyringia
			Thenea
			Thrombus
		Spirophorida	*Cinachyra*
			Tetilla
		Lithistida	*Corallistes*
			Monanthus
			Petromica
		Hadromerida	*Suberites*
			Polymastia
			Stylocordyla
			Spirastrella
			Cliona
			Alectona
			Thoosa
			Timea
			Tethya

Class	Subclass	Order	Genus
Demospongiae			
			Aaptos
			Chondrosia
			Sigmoscepterella
			Latrunculia
			Spheciospongia
		Axinellida	*Axinella*
			Homaxinella
			Phakellia
			Raspailia
			Myrmekioderma
			Ceratopsion
			Trachycladus
			Agelas
	Ceractinomorpha	Halichondrida	*Halichondria*
			Ciocalypta
			Hymeniacidon
		Poecilosclerida	*Mycale*
			Hamacantha
			Crella
			Lissodendoryx
			Coelocarteria
			Ophlitaspongia
			Microciona
			Clathria
			Neofibularia
			Crambe
			Ectyomyxilla
			Iophon
			Tedania
			Hamigera
			Tetrapocillon
		Haplosclerida	*Haliclona*
			Reniera
			Adocia
			Strongylophora
			Orina
			Callyspongia
			Dactylia
			Spongilla
			Corvomeyenia
			Ephydatia
			Gellius
			Gelliodes
			Xestospongia

Class	Subclass	Order	Genus
Demospongiae		Dictyoceratida	*Spongia*
			Hippospongia
			Phyllospongia
			Fasciospongia
			Cacospongia
			Dysidea
			Stelospongia
			Polyfibrospongia
			Ircinia
			Sarcotragus
		Dendroceratida	*Darwinella*
			Dendrilla
			Aplysilla
			Halisarca
			Bajalus
		Verongida	*Verongia*
			Psammaplysilla
			Pseudoceratina
			Ianthella
Sclerospongiae		Ceratoporellida	*Ceratoporella*
			Astrosclera
			Merlia
			Stromatospongia
			Goreauiella
		Tabulospongida	*Acanthochaetetes*

1 The basic sponge: its organization and operation

1.1 Structure and level of organization

Sponges are complex, sedentary filter-feeding organisms. They differ from all other groups of invertebrates which occupy similar ecological niches in that they maintain an almost protozoan independence for their constituent cells, while at the same time ensuring that the entire cell mass pumps sufficient water to effect all essential exchanges. The life of a sponge centres around pumping a high volume of water through the tissues at low pressure, and the organization of the body is around the system of pores, ostia, canals and chambers which conduct the water current from the inhalant sponge surface to the exhalant apertures, the oscules.

Specialized flagellated cells (choanocytes), each with a collar of cytoplasmic tentacles, are the effectors which drive the water current (Fig. 1.1). These choanocytes are arranged as a single-layered epithelium which can be simple

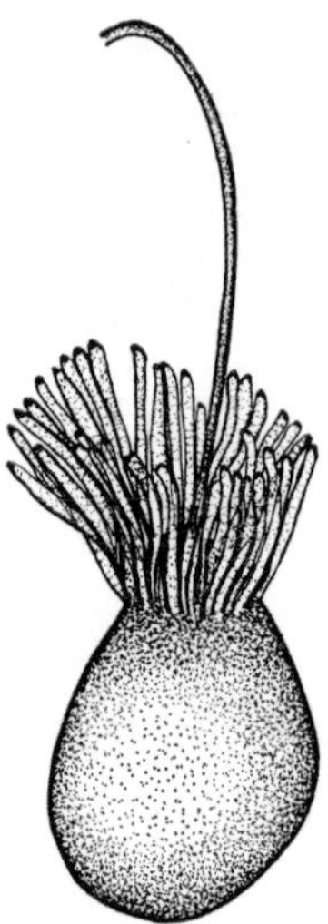

Fig. 1.1 An isolated choanocyte with long flagellum, apical collar of cytoplasmic tentacles and rounded cell body. (Redrawn from Rasmont 1959.)

and continuous as in *Leucosolenia*, folded and continuous as in *Ascandra*, folded and discontinuous as in *Sycon*, or greatly subdivided, the condition which exists in most sponges (Figs. 1.2, 1.3). A further epithelial layer, also one cell deep and composed of pinacocytes, separates the sponge from the external environment. This pinacoderm can be simple, or it can fold to line the internal cavities of the aquiferous or water circulatory system. Pinacocytes are distinguished by their position as exo-, endo- or basopinacocytes, and show functional and structural diversification which never occurs among the choanocytes.

Between the pinacoderm and the choanoderm a connective tissue (the mesohyl), is interposed; increase in size and complexity in the evolution of sponges has been dependent upon increase in complexity of this mesohyl. Here the skeletal components, organic and inorganic, are secreted and organized. As these become numerous and diverse a more intricate and efficient aquiferous system can be set up and the diversification of cell types becomes more pronounced.

The simplest type of sponge organization is that found in primitive Calcarea such as *Leucosolenia* and *Clathrina*. In these sponges the body remains a simple tubular unit, with thin walls enclosing a central cavity which opens apically by a single osculum. The pinacoderm is interrupted by specialized pinacocytes which, in development, elongate and roll to enclose a cylindrical canal. These cells traverse the thin mesohyl and pierce the choanoderm between the bases of adjacent choanocytes, thus placing the external medium in direct communication with the central choanocyte layer. Cells enclosing a pore canal are termed porocytes and, in *Leucosolenia*, they are known to make contact with the base of the pinacocytes and choanocytes to which they are adjacent (Jones 1966). This is probably a general condition in simple sponges, for it ensures that a continuous epithelial surface separates the water current from the mesohyl. A simple unfolded layer of choanocytes lines the internal cavity (Figs. 1.2*a*, 1.3*b*). Construction of this type is termed asconoid, it most frequently occurs in the olynthus (Fig. 1.4), the post-settlement stage in calcareous sponges, and is retained in this simple form in adult sponges in only *Leucosolenia* and *Clathrina*.

From such an asconoid unit, folding of both pinacoderm and choanoderm produces a syconoid type of organization and here several levels can be recognized. Essentially, the inner choanoderm surface is amplified to line a series of projections which extend radially outward from the central cavity. The pinacoderm is folded outward to invest the projections. Choanocytes are now restricted to lining the choanocyte chambers each of which opens to the central, atrial cavity by a wide aperture, the apopyle. In this elementary syconoid condition the mesohyl has undergone relatively little thickening and thus inward flow of the water current can still be effected by way of porocytes, several of which open to each choanocyte chamber (Fig. 1.2*b*). If now the mesohyl thickens and the outer ends of the radial projections fuse to produce a smooth outer surface, the result is to produce a cortical region in which additional, often distinctive, skeletal elements can be deposited. The separation of inhalant surface and

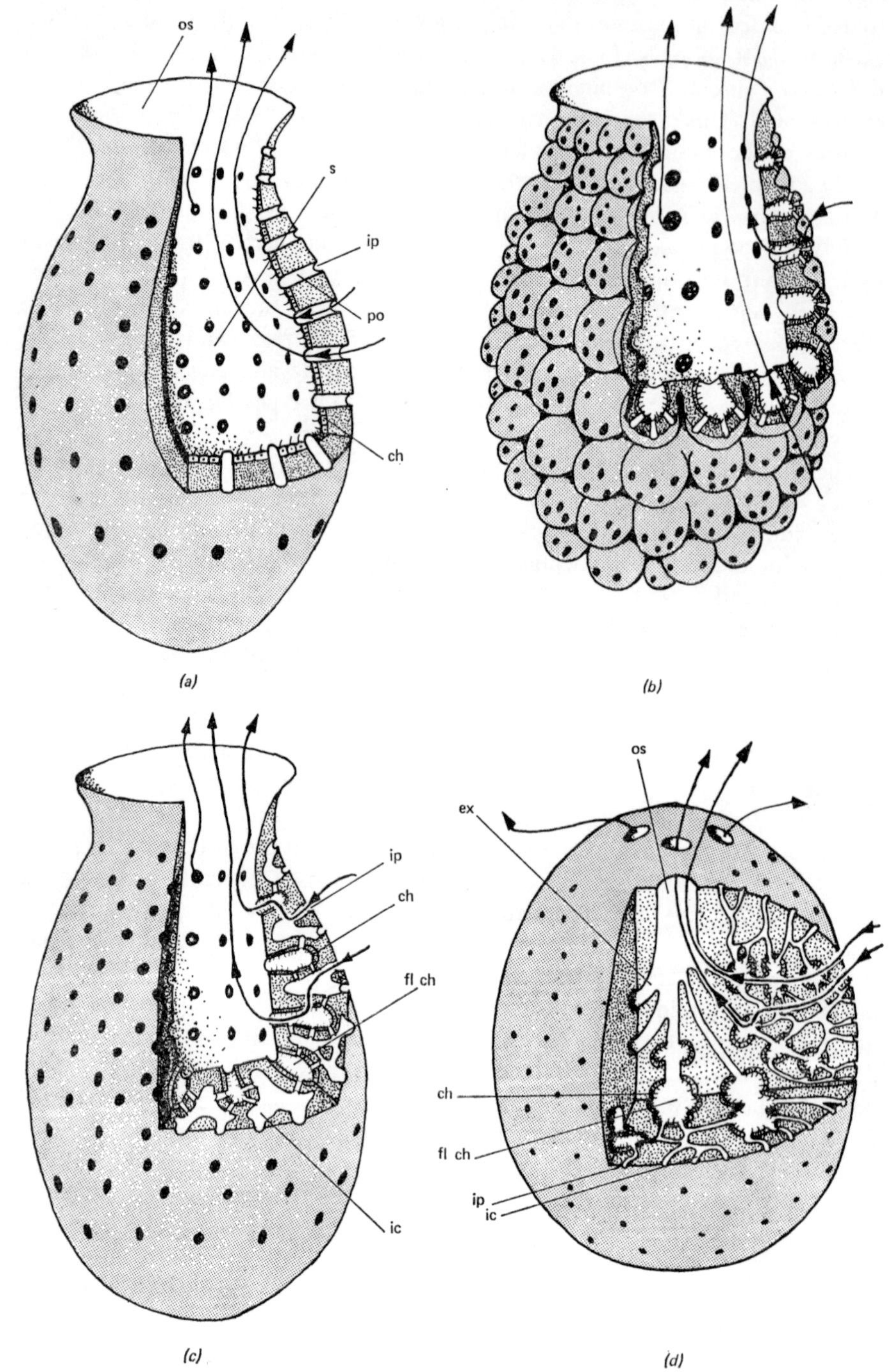

os
s
ip
po
ch
(a)
(b)
ip
ch
fl ch
ic
(c)
os
ex
ch
fl ch
ip
ic
(d)

Fig. 1.2 A diagrammatic representation of the different levels of sponge organization. (*a*) The simple vase with unfolded epithelia (ascon); (*b*) the first stages of folding (sycon); (*c*) a more advanced syconoid condition; (*d*) the most complex condition with isolated choanocyte chambers, complex canals and thick mesohyl (leucon). Arrows mark the direction of the water current.

os, osculum; s, spongocoel; ip, inhalant pore; po, porocyte canal; ch, choanoderm; flch, choanocyte chamber; ic, inhalant canal; ex, exhalant canal.

(Redrawn from Baer and Owre, *The Free Living Lower Invertebrates*, Macmillan 1968.)

choanocyte chambers by the interposition of the cortex necessitates the development of an inhalant system. Most frequently this is a system of superficial ostia opening into inhalant lacunae, lined by pinacocytes, which then open by way of porocytes, or pores surrounded by several pinacocytes, prosopyles, to the choanocyte chambers (Fig. 1.2*c*). Choanocyte chambers become elongated radially in syconoid forms, and often in the early literature were termed radial canals. It is more precise, however, to refer to any cavity lined by choanocytes as a choanocyte chamber, thus avoiding confusion between components of inhalant and exhalant canal systems. If further mesohyl thickening and skeletal deposition takes place between the atrial ends of the choanocyte chambers and the point at which the apopyles open to the atrium, then a system of exhalant canals is developed. The atrial cavity is lined by a pinacoderm and opens by a single osculum. Organization of syconoid type occurs in many Calcarea (e.g. *Sycon*) (Figs 1.2*b*, *c*, 1.3*c*, *d*).

Further folding of the choanoderm is accompanied by subdivision of the flagellated surface into discrete spherical or oval chambers which are the typical flagellated, or choanocyte, chambers. This feature characterizes the leuconoid grade of organization. Choanocyte chambers cluster in groups in the thickened mesohyl, and each chamber is serviced by two or more inhalant canals which open through prosopyles and is drained by a single apopyle which leads to an excurrent canal. The excurrent canals coalesce to form larger channels and these converge toward the exhalant apertures (Figs 1.2*d*, 1.3*e*, *f*). This pattern of inhalant and exhalant canals with dispersed choanocyte chambers can become extremely complex, and the approximation to a radially symmetrical morphology, which can be recognized in asconoid and syconoid forms, is no longer evident. A great increase in the volume of the mesohyl, and a diversification of the cellular and skeletal elements it contains, accompanies the development of a leuconoid aquiferous system. This allows the development of very large sponges with high pumping efficiency. Leuconoid construction is encountered in most Calcarea and in all other sponges.

The recognition of levels of organization encountered in sponges allows one to convey in a single word the degree of subdivision of the choanoderm, folding of the pinacoderm, and increase in volume and complexity of the mesohyl. It

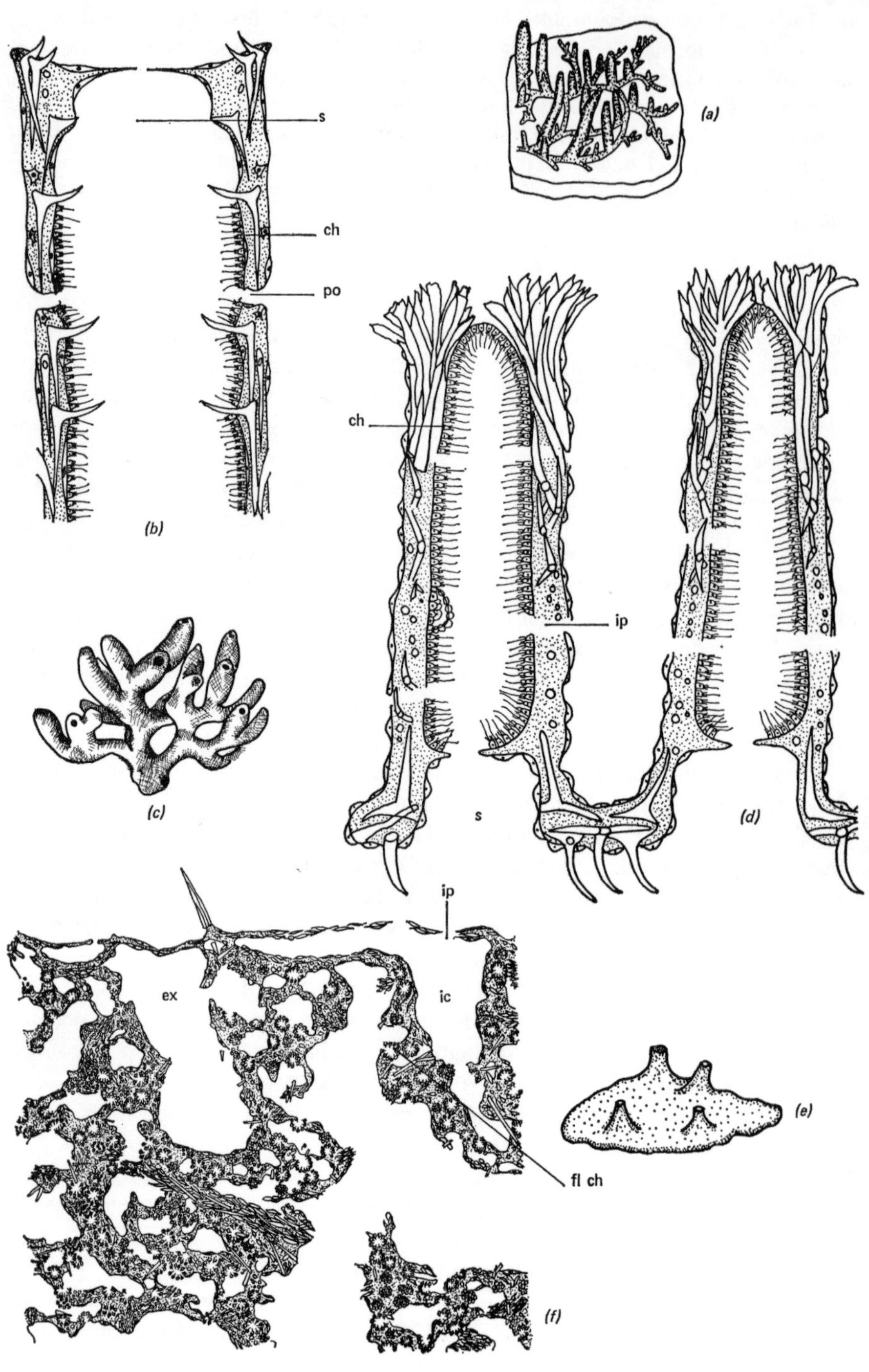
(a)
s
ch
po
(b)
(c)
ch
ip
s
(d)
ip
ex
ic
(e)
fl ch
(f)

Fig. 1.3 The three basic types of sponge organization drawn to show cellular disposition in relation to actual body shape. (*a*) An individual *Leucosolenia*; (*b*) a vertical section of one oscular tube of *Leucosolenia*; (*c*) an individual of *Sycon*; (*d*) a transverse section through one oscular tube of *Sycon*; (*e*) an individual *Haliclona*; (*f*) a section of *Haliclona* taken at right angles to the surface.
s, spongocoel; ip, inhalant pore; po, porocyte canal; ch, choanoderm; flch, choanocyte chamber; ic, inhalant canal; ex, exhalant canal.
((*a*) Redrawn from Minchin 1900; (*b*) redrawn from Jones 1966; (*d*) redrawn from Bergquist 1972; (*f*) redrawn from Hartman 1958*a*.)

does not imply an evolutionary or developmental sequence. Leuconoid construction clearly has been achieved independently in many lines during sponge evolution; it is not necessary that all sponges pass, either in phylogeny or ontogeny, through asconoid and syconoid stages.

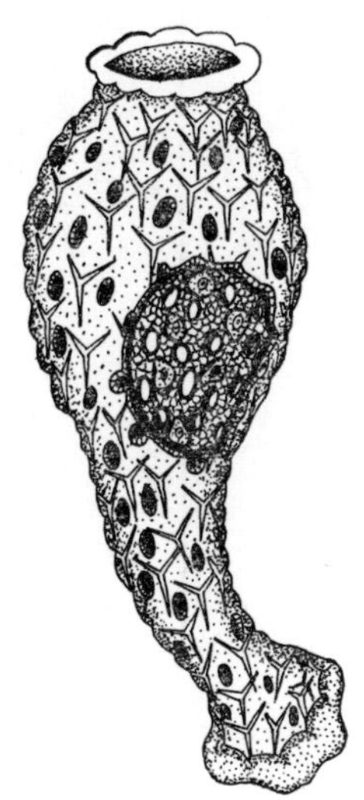

Fig. 1.4 Diagrammatic representation of the olynthus, an asconoid state which occurs after settlement in calcareous sponges. (Redrawn from Haeckel 1872.)

Table 1.1 *Characteristics of the major organizational levels recognized in Porifera.*

Grade of organization	Occurrence	Pinacoderm	Condition of the mesohyl	Choanoderm	Path of water current
Asconoid	Olynthus *Leucosolenia* *Clathrina*	Simple. Inhalant current passing through canals bounded by single porocytes	Thin. Spicule differentiation from oscule towards base rather than radial	Simple, unfolded lining a single central cavity	Porocyte → choanoderm → osculum
Syconoid	Some *Sycon* species e.g. *Sycon conifera*	Folded. Inhalant current passing through canals bounded by single porocytes	Thin. Some radial spicule differentiation	Folded outwards into a series of radial lobes. Choanocytes restricted to lobes which open by apopyles	Porocyte → choanocyte chambers → apopyles → atrial cavity → osculum
	Some *Sycon* species e.g. *Sycon gelatinosum*	Folded. Superficial osita opening to lacunae or inhalant canals bounded by pinacocytes. Porocytes where they occur are between lacunae and choanoderm	Thickened, with a cortex and regional spicule differentiation	Folded into elongated chambers opening by a narrowed apopyle to the exhalant canal or atrial cavity	↗ lacunae → porocytes Ostia ↘ canals → prosopyles → choancocyte chambers → ↗ exhalant canals apopyles ↘ atrium → osculum
Leuconoid	Some Calcarea e.g. *Leucetta* all Demospongiae	Folded. Ostia only, no porocytes. Pinacocytes line all inhalant and exhalant canals	Greatly thickened with diversified skeletal and cellular elements	Restricted to spherical or oval chambers which are dispersed in the mesohyl	Ostia → inhalant canals → prosopyles → choanocyte chambers → apopyles → exhalant canals → osculum

1.1.1 Hexactinellida: a special case

The glass sponges, or Hexactinellida, differ considerably from Calcarea and Demospongiae in basic organization. Many assumptions now made about their cellular organization will certainly change when electron microsopic studies are completed, but we must note here the main structural features of hexactinellids as interpreted at present. Most illustrations in the literature stem from the work of Schulze (1800, 1887) and Ijima (1901). The only detailed study after this date is that by Okada (1928) which concentrates on development and which provides our total knowledge on that topic.

The body of hexactinellid sponges, more frequently than in any other group appears, externally, to be radially symmetrical, various modifications of a cylindrical tubular construction are common (Fig. 1.5). This appearance is reinforced in many forms by the fusion of spicules to establish a permanent framework, as for example in *Farrea* (Fig. 1.6*b*).

Cellular material is distributed sparsely from the outer (inhalant) surface of the body wall to the inner (exhalant) surface and is in the form of a trabecular net stretched across intercommunicating cavities (Fig. 1.6). The choanocyte chambers are supported within the trabecular network and are arranged as a series of oval spaces, open toward the central cavity, from which they are separated by an extension of the net (Fig. 1.6*a*, *c*). There is no obvious equivalent of a pinacoderm in hexactinellids. A dermal membrane is present but it is extremely thin and no continuous cellular structure supports it. Ostia, which must exist to allow water to enter, are simply depicted as holes in a film-like membrane. Such a structure would not permit any control over water ingress, and while it is preferable to believe that pinacocytes will eventually be discerned in hexactinellids, it is conceivable that, in the stable deep-water environment in which most Hexactinellida are found, no control is required. Schulze (1880, 1887) has argued that nuclei of pinacocyte type do occur in a superficial position, but Ijima (1901) has criticized this assumption in a careful analysis of euplectellid structure and has shown that the cells in question were not superficial or continuous and were probably what are now referred to as archaeocytes. The controversy has remained at this point over the intervening period. Ijima's interpretation is generally accepted and his theory that absence of a pinacoderm is possible in hexactinellids because of the absence of the non-cellular component of the mesohyl, the matrix, has never really been discussed. The mesohyl in Hexactinellida is thought to be represented only by the trabecular network and the siliceous skeleton.

There is another possibility. Recent work on Demospongiae has shown that the superficial flattened portions of the pinacocytes can be resorbed locally and reformed later, ostia do not appear in the areas which are undergoing reorganization (Vacelet 1971*a*). It is possible that the pinacoderm of Hexactinellida is in a condition of considerable flux and that the formation and disappearance of incurrent apertures is a constant process. This would account for the difficulty encountered in identifying a continuous pinacoderm in histological preparations.

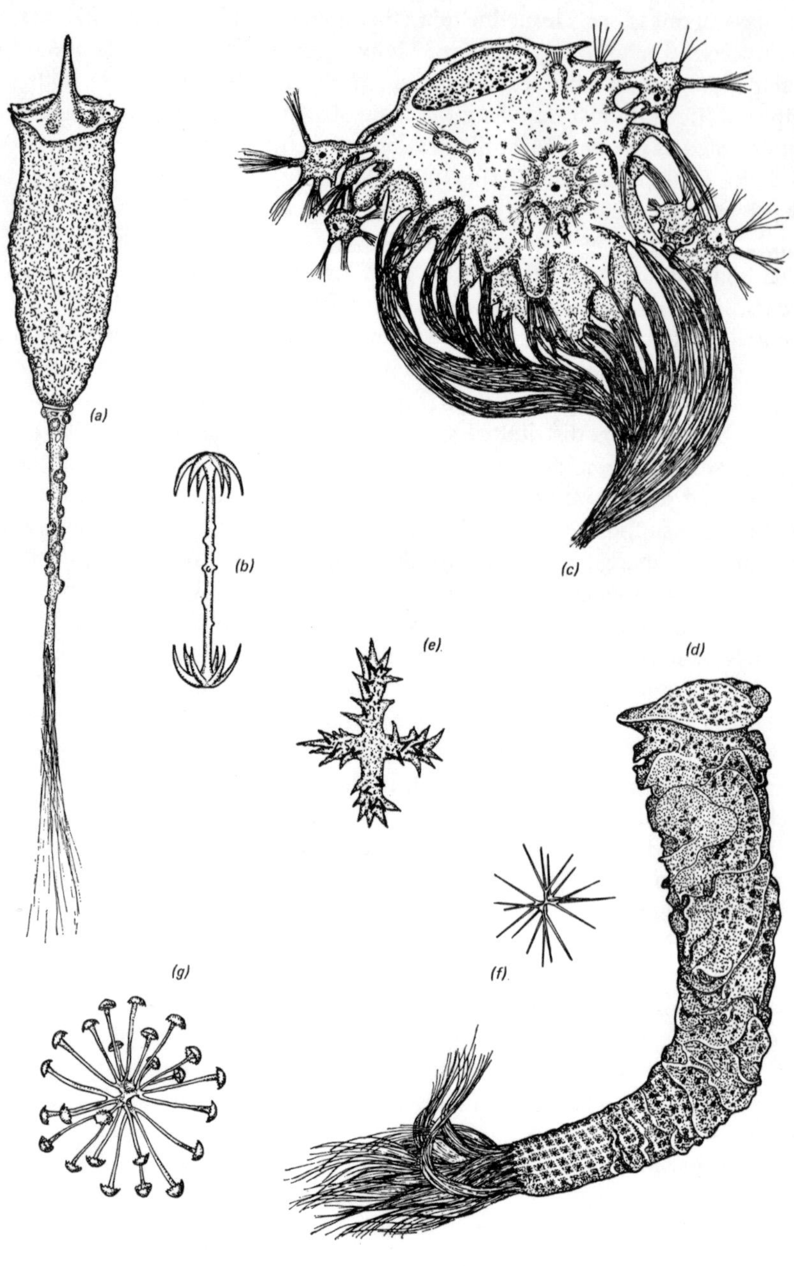

(a)
(b)
(c)
(d)
(e)
(f)
(g)

Fig. 1.5 Body form and microsclere spiculation of some typical Hexactinellida. In all three examples an approach to external radial symmetry is evident. This tendency is very common in deep-water sponges belonging to both the Hexactinellida and the Demospongiae.
(*a*) *Hyalonema thompsoni.* A representative of the subclass Amphidiscophora.
(*b*) The amphidisc microsclere which is found in all members of the Amphidiscophora.
(*c*) *Lophocalyx philippinensis.* A large, vase-shaped hexactinellid belonging to the subclass Hexasterophora. This particular species produces abundant surface buds.
(*d*) *Euplectella aspergillum.* The 'venus flower basket' sponge. Another example of the Hexasterophora. The arrangement of megasclere skeletal elements into a regular lattice and the sieve plate over the osculum are well shown in this species.
(*e*) Tetractinal basal spicule of *Hyalonema thompsoni.*
(*f*) An oxyhexaster, one of the simple asterose microsclere types found in *Euplectella.*
(*g*) A discospiraster, one of the more elaborate microscleres which occur in the family Euplectellidae.
(All redrawn from Schulze 1887.)

Usually, five regions are recognized in a section of the body wall of a hexactinellid (Figs 1.6*a*, *e*). The dermal membrane, which is perforated by ostia which allow entrance of water into the intertrabecular lacunae. A subdermal trabecular network, which is continuous with the dermal membrane; this is dense toward the periphery and cavernous toward the interior, where large subdermal spaces are developed. These cavities are in contact with the outermost choanocyte chambers, at which point the trabecular extensions are very thin and choanocytes are almost directly exposed to the external medium. Distinct inhalant canals lead to more deeply situated choanocyte chambers in species where the body wall is thick and the choanocyte layer folded. The choanocyte layer itself is peculiar in hexactinellids. It is composed of choanocytes, the bases of which are continuous to form a reticular membrane perforated with apertures, the prosopyles, through which water passes from the subdermal trabecular lacunae into the choanocyte chamber. Each chamber opens by way of a large apopyle to the ill-defined excurrent channels, which then traverse the inner trabecular network and open to the central atrial cavity. There is a single, terminal osculum at the apex of the atrial cavity and this, in some species, is covered by a complex, spicule-supported sieve plate (Fig. 1.5*d*).

The trabecular net is considered to be formed by the union of long pseudopodia produced by collencytes and archaeocytes. Cells which compare with the archaeocytes of other Porifera do occur in the trabecular net and reproductive elements move freely through the network.

Some workers term this almost syncytial structure an ectomesenchyme; however, it is worthless to speculate on the homologies of hexactinellid body layers until more is known of their structure and development. Similarly, it

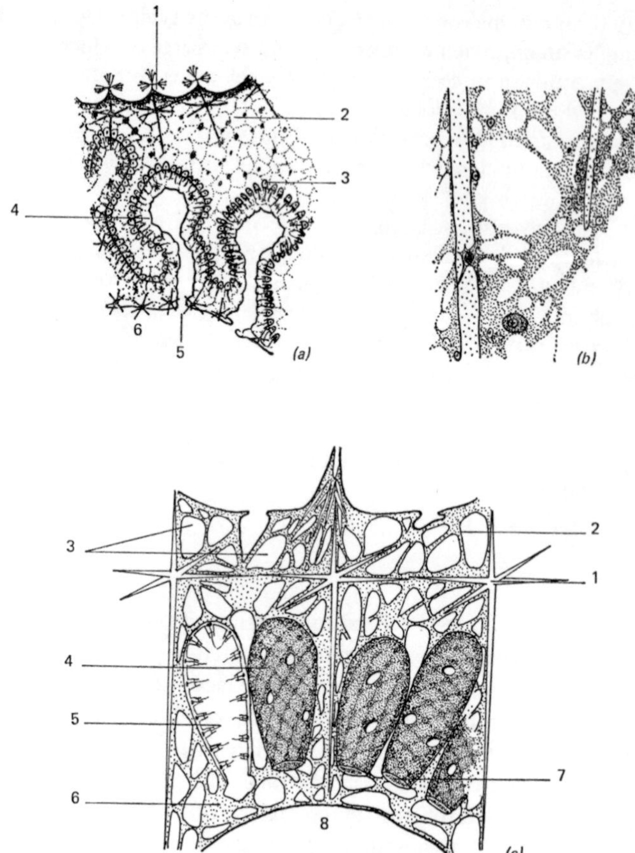

Fig. 1.6 The internal organization of Hexactinellida.
(*a*) Transverse section through the body wall of *Euplectella*.
(1) Dermal layer with specialized dermal spicules. (2) Subdermal trabecular net. (3) Choanocyte chamber with lining of choanocytes, note that the flagella extend into the inner trabecular network. (4) Inner trabecular network. (5) Apopyle. (6) Atrial cavity lined by gastral spicules.
(Redrawn from Schulze 1887.)
(*b*) A diagrammatic representation of a portion of the trabecular network of *Farrea*. Cells are sparse; notable concentrations are found only around spicules. (Redrawn from Okada 1928).
(*c*) A diagrammatic representation of a section perpendicular to the surface of *Euplectella*. The section was taken through a ridge (see Fig. 5*d*) where the extent of the trabecular layer is reduced.
(1) Large hexactinal megascleres. (2) Subdermal trabecular network. (3) Subdermal lacunae. (4) Choanocyte chambers with inhalant apertures, the prosopyles, shown. (5) Choanocyte chamber in section. (6) Inner trabecular network. (7) Apopyle. (8) Atrial cavity.
(Redrawn from Schulze 1887.)

is not possible to encompass hexactinellid structure within the ascon–sycon–leucon progression which is useful when considering other Porifera.

1.2 The physiology of sponges

Having outlined in structural terms how the simple components of the sponge body can be disposed to allow increase in size and, as shall be seen later, increase in efficiency, we will now define the function of the apertures, the cells and the canals in generating and utilizing the water current. We can then examine very briefly the major organizational strategies sponges have employed as they have become large, more efficient and even active organisms.

1.2.1 Cell mobility

A sponge is able to function with the simple morphological equipment we have described, by retaining for all of its cells a high degree of mobility and for many a low degree of canalization. Some sponge cells differentiate irreversibly, such as those committed to reproduction, or specialized for secretory activity and skeleton production; other cells retain differentiative capacity and can change their function. It has been recognized for many years that the cells of the sponge mesohyl are freely mobile and that choanocytes may move from the choanocyte layer to become amoeboid or, more frequently, to become sperm. However, the mobility of all cells, including pinacocytes, has been demonstrated graphically by recent studies using time-lapse cinematography. It is still true to say that cells of the adult pinacoderm and choanoderm are more stable in terms of position and state of differentiation than most mesohyl elements, but all cells can become amoeboid and the whole structure is a continuously mobile system. The inter-relationships between various differentiated cell states and the functions ascribed to each will be dealt with in the next chapter, but at this point it is significant to note the importance of cell mobility and change of function in the life of a sponge.

1.2.2 Nutrition: ingestion, digestion and excretion

Sponges are unselective particle feeders and, in effect, the arrangement of inhalant ostia, canals, prosopyles, choanocyte collar tentacles and inter-tentacular mucous reticulum places a series of sieves of decreasing mesh size in the path of the water current. The diameter which can be attained by expanded ostia in the pinacoderm, or by porocytes in simple sponges, sets the upper size limit, effectively 50 μm, of the particles which a sponge can filter from its environment. There are reports of direct phagocytosis of particles larger than 50 μm by the exopinacocytes of fresh-water sponges, but most deal with uptake of non-nutritive particles under laboratory conditions. Recently, Harrison (1972) has established that the basopinacocytes of *Corvomeyenia*, a fresh-water sponge, actively ingest and digest bacteria. These bacteria were of a size which could also have been filtered by the aquiferous system. It remains to be demonstrated that surface phagocytosis of large particles occurs in fresh-water sponges under natural conditions, or that it occurs at all in marine sponges.

Inside the sponge particle capture operates at two levels, as evidenced both by direct observation and also by comparison of the particle composition of inhalant and exhalant streams, where a clear bimodal pattern of particle retention has been revealed.

Mobile archaeocytes which move close to the lining of the inhalant canals operate the primary capture system, which deals unselectively with particles in the 2–5 μm size range. Phagocytosis can be direct, as particles come into contact with the canal wall, or can occur as particles become trapped by the progressively narrowing canal lumen near the prosopyles. The dual purposes of feeding and of preventing canal occlusion are served by phagocytosing archaeocytes. Naturally, non-selective phagocytosis of this type leads to the incorporation of much non-nutritive material in vacuoles, and necessitates constant elimination of waste material.

There is no definite evidence available on how a finite archaeocyte population in the mesohyl deals with both phagocytosis and particle elimination at a sufficient rate to operate efficiently at the observed rates of water pumping.

There are three possibilities. Archaeocytes can cycle constantly from inhalant to exhalant systems, i.e. phagocytose, digest and excrete. If this occurs, and if the archaeocyte proportion of the cell population is constant for any particular species, we should expect to observe differences in efficiency of particle retention between species, and should be able to relate these differences to the architecture of the sponge, which of course governs the rapidity with which a given cell can complete a cycle of ingestion, digestion, excretion and then return to phagocytose again.

Alternatively, it is possible that architecture of a species, primarily its mesohyl density, is not of great importance. Sponges can engage in major reorganization of the course that inhalant and exhalant channels take through the mesohyl, with the effect that inhalant and exhalant routes can be brought closer together, thus expediting particle elimination. If canal reorganization involves the entire sponge, it should be detectable as a period of reduced or halted water flow. There is direct evidence from a time-lapse cinematographic study of spongillids that canal migration is common in young sponges (Kilian 1964). However such sponges, observed between microscope slides in the laboratory, are subject to unusual hydrodynamic stresses, and being very young may not have a full complement of archaeocytes. Nevertheless, this mechanism remains a real possibility as an adjunct to efficient feeding, since we know from observation that canals do appear and disappear frequently in all sponges, and we do not know whether this relates to cell activity cycles or simply to increase in volume as a result of growth.

Lastly, it is possible that at times when canal occlusion occurs, as for example after a storm, the normal archaeocyte population is unable to complete the cleaning task rapidly enough. At these times other cells such as collencytes and pinacocytes may engage in phagocytosis from canals, certainly all of these cell types have been observed to possess phagocytic vacuoles. They could also supplement archaeocyte activity under normal feeding conditions.

At present there are many assumptions and few facts, the indications are that all three processes can operate, but will differ in relative importance between species and in different habitats. The implications of different feeding strategies for the ecology of Demospongiae will be discussed later.

Secondary particle capture operates at the level of the choanocyte collar. After entry to the chambers by way of the prosopyles, particles are trapped between the collar tentacles and on the fibrillar reticulum or mucous net which extends between collar tentacles. It is certainly here that particles in the bacterial size range, 0.1–1.5 μm, are ingested. Although the diameters of inhalant ostia, canals and prosopyles vary from sponge to sponge and thus particulate material of a range of sizes can be admitted to the choanocyte chambers, the spacing of choanocyte collar tentacles is near constant at 0.1–0.2 μm in all species where it has been observed.

Direct observation of particle capture by the choanocytes has been made frequently. Electron microscopy reveals that active microendocytosis takes place at the choanocyte surface. It has been estimated that in Demospongiae 81 % of the particulate organic carbon intake is captured by the choanocyte system; certainly all sponges retain bacteria at high efficiency, around 96 % capture in Demospongiae, and this takes place in great part at the choanocyte level.

Although there is no question that choanocytes phagocytose a high proportion of the sponge diet there is some uncertainty as to the cellular site of digestion and assimilation. In the case of the primary capture system, digestion takes place in the vacuoles formed at the time of phagocytosis, the phagosomes, Thus, the extent to which particular, potentially nutritive particles (such as armoured algal cells) are digested is governed by the time taken for the ingesting cell to migrate from inhalant to exhalant system. In the case of choanocytes, however, ingested material is quickly processed to the base of the cell and transferred to amoeboid mesohyl cells, chiefly the archaeocytes, in which digestion then proceeds. Pourbaix (1933) and Kilian (1952) fed sponges with non-nutritive particles such as carmine and indian ink and established that these were captured by choanocytes and passed on to archaeocytes.

In an elegant recent study, Schmidt (1970) fed fluorescent antibody-labelled *Escherichia coli* to *Ephydatia fluviatilis* and was able, by monitoring the occurrence of fluorescent material, to place a time estimate on the transfer of phagocytosed material. Thirty minutes after feeding, labelled bacteria could be detected in vacuoles near the base of the choanocytes; electron micrographs revealed many small phagosomes, 0.1–0.5 μm in diameter, containing recognizable bacteria congregated near the choanocyte apex. Passage of fluorescent material toward the mesohyl commenced after 30 min and phagosomes were accepted from choanocytes by archaeocytes, microgranular cells and collencytes. After 3 hours bacteria were dispersed in the mesohyl inside these cells, but remained unaltered in appearance. Four to five hours after feeding, the dispersed fluorescent inclusions changed in character, and the bacterial shape was no longer apparent. The distribution of fluorescence at this time was very general throughout the sponge, congregated basally, around oscules, and in the pinacoderm. Some of this could have resulted from aggregates of bacteria phagocytosed at the pre-choanocyte stage, but the important point was that no fluorescence remained in the choanocytes. After 24 hours all fluorescent material was concentrated in large heterogeneous phagosomes within archaeocytes which then discharged into exhalant canals. All fluorescent material had left the sponge after 48 hours. This work indicates very clearly that no substantial digestion takes place in choanocytes and supports earlier observations (Van Weel 1949) that particles are not retained in choanocytes for more than 3 hours.

A series of observations often cited to support the reverse hypothesis, that digestion occurs in choanocytes to a greater extent than in archaeocytes, was made by Agrell (1951). He used density gradient centrifugation to obtain two fractions from cell suspensions of *Halichondria panicea*, a marine demosponge. One fraction was claimed to consist of 90 % choanocytes, the other of 80 % archaeocytes and 20 % choanocytes. Both fractions were assayed, after freeze-thawing, for protoelytic, lipolytic and amylytic activity. The conclusion was, using cell volume as an index, that proteolytic activity at pH 4.4 is 8 times higher in choanocytes than in archaeocytes, at pH 6.8, 6 times higher; lipolytic activity was similarly 15 times higher and that of amylase 8.5 times higher.

In determining the cell composition of his two fractions, Agrell grouped all cells other than choanocytes as archaeocytes, 48 % of the total cell population; however, from his descriptions and his references to other authors, it is certain that, at a maximum, only 40 % of his archaeocyte fraction would fit the modern definition of archaeocyte (see Chapter 2). With the more sophisticated techniques for cell separation and enzyme assays now available it would be profitable to repeat similar biochemical work in an effort to define the enzymatic activities of particular cell types. However, as it stands, this hypothesis of choanocyte digestive activity in Demospongiae lacks experimental support.

The discovery of the nature and role of lysosomes and the development of histochemical techniques for locating lysosomal acid phosphatase at the ultrastructural level should allow a further approach to the problem of locating digestion as distinct from ingestion in sponge cells. Thiney (1972) has found acid phosphatase activity in five cell types of *Hippospongia communis:* endopinacocytes, exopinacocytes, archaeocytes, myocytes, and spherulous cells. She did not study choanocytes, but the wide dispersion of activity, associated in all cases with lysosomes, suggests that all cells will show this activity but that it is not necessarily always related to a digestive process. Lysosomes have many auto-digestive roles in the cell, and this could account for the acid phosphatase activity which is present in myocytes and spherulous cells.

There are some Demospongiae (*Plakina, Oscarella*) where, in the adult, the population of mesohyl cells is greatly reduced, the sponge is essentially just pinacocyte, skeleton and choanocyte. In this case it is certain that choanocytes will have additional digestive and assimilative roles. Volkonsky (1930) has produced good evidence that in some Calcarea, where the mesohyl cell populations are also sparse, choanocytes have a digestive role.

In the final analysis it is most probable that no general rule will apply, different groups in different habitats will vary, certainly that is the inference one must take from the information to hand. There is a propulsive and prehensive role which choanocytes always exhibit; digestion, it appears, can take place in many cell types.

While considering capture of nutrients, we should note that it is likely that sponges are able to supplement their particle feeding by absorbing dissolved or colloidal nutrients directly from the surrounding water. It is never possible to be sure that this does not occur, but efforts to sustain fresh-water sponges on a diet of dissolved nutrients have failed. Also, in the three cases where carbon budgets which take no account of dissolved nutrients have been calculated, there is no substantial deficit in two cases, and an endemic bacterial population which complicates the third. However, Schmidt (1970) was able to observe the uptake of fluorescent-labelled casein and rabbit serum protein from the medium by *Ephydatia fluviatilis*, and she determined that pinocytosis takes place preferentially at the choanocyte surface and is followed by transfer of the material to archaeocytes.

Sponges do not produce any extracellular proteolytic enzymes; digestion is strictly intracellular. Regardless of the site of particle capture, once assimilation is complete expended phagosomes are discharged, either to the exhalant system, or directly to the exterior.

We have seen that two levels of particle capture operate in sponges and that together these are able certainly to deal with particles from 50 μm, a general figure for ostial diameter, down to 0.1 μm, the spacing of the choanocyte collar tentacles. Reports of direct surface phagocytosis by pinacocytes are likely to be correct thus extending upward the size of particle which is available to a sponge. At this level, however, cell size itself would soon limit the dimensions of particles which could be phagocytosed.

The demonstration that uptake of protein molecules from solution takes place by choanocyte pinocytosis and of the existence of the mucous mesh between collar tentacles, suggest that filterers of the sponge type may be able to utilize material from 50 μm down to soluble single molecules.

All literature reporting artificial feeding of sponges is unsatisfactory in one way or another, but mainly because materials used were non-nutritive or because the sponges were functioning under adverse laboratory conditions. Certainly we cannot glean from such reports much information on the natural diet of a sponge. A summary of all reports, which relate mainly to fresh-water sponges and Calcarea, is that sponges are unselective particle feeders able to capture a wide size range of particles and able to retain particles of bacterial size efficiently.

This observation is of little use if we wish to assess the ecological role of sponges as major components of marine biotopes. This surely must be a major aim of studies on feeding. We understand, at least in outline, the mechanisms involved; we know nothing of the exchanges of organic material taking place between the sponge and its food source.

In order to approach this problem it is necessary to provide answers to three questions. What food is available in the environment? What do sponges filter from the available supply? Of the material filtered, what is utilized and what is excreted; that is, is there any evidence of particle selection?

There are great technical difficulties in the way of obtaining this information, the primary one being that the huge volumes of water pumped by Demospongiae make them difficult to sustain under experimental conditions hence the observations must be made in the field. An added limitation is that the composition of available particulate organic carbon in the sea water varies from habitat to habitat, a diet determined for a sponge in one location may differ markedly from the diet available to sponges in another habitat. Recently, however, a study of this type has been completed (Reiswig 1971*b*) and drawing on this data, we can provide some answers to the above questions.

(a) What food is available?

The habitat studied was the fore-reef slope and shallow enclosed waters of Discovery Bay, North Coast, Jamaica, and observations apply in detail only to that area or to areas where the spectrum of particulate material is closely comparable. Since, however, quantitative analyses of planktonic biotopes are much more numerous and sophisticated than existing analyses of benthic and intertidal habitats, this is not as restrictive as it may seem. The techniques employed in the study and the statistical procedures need not be examined here; suffice to say the methods were rigorous and are well described in the literature.

Exchanges of particulate organic material in undisturbed individuals of three species of Demospongiae were monitored by collecting water at the same time just above the inhalant surface and from the oscular stream. Samples then were analysed for particle composition by direct microscopy, and for total particulate organic carbon (POC) by standard chemical means. Comparison of these pairs of inhalant/exhalant samples provided information on the removal of planktonic particles by type and size, and thus allowed an assessment of the natural diet of the species studied.

The major fractions recognized by direct microscopic analysis are as listed in Table 1.2 where their individual proportions in the plankton by number, volume and weight of carbon are given.

The organic carbon content of these fractions was determined for all samples and estimates of available carbon, particle concentration and particle volume were made. When figures obtained from these analyses were compared with total carbon estimates obtained by chemical means, only 12–14 % of the available carbon proved to derive from microscopically visible particles. This suggests that the major portion of the available particulate carbon in Jamaican water,

Table 1.2 *Composition of particulate carbon in inhalant water (from data in Reiswig 1971b).*

Fraction	*Size* (μm)	(1) Particle concentration (no. ml^{-1})		(2) Particle volume ($10^3 \mu m^3\ ml^{-1}$)		(3) Calculated organic carbon ($mg\ m^{-3}$)		(4) Chemically determined carbon ($mg\ m^{-3}$)		(5) Unresolvable particulate carbon (i.e. col. 4 − col. 3)	
		Bay	*Reef*	*Bay*	*Reef*	*Bay*	*Reef*	*Bay*	*Reef*	*Bay*	*Reef*
Bacteria	0.3–0.8	114000	39740	5.7	2	0.6	0.2				
Unarmoured cells	2–50 diameter	275	227	43	28	7	4.5				
Armoured cells (fungi, diatoms, dinoflagellates, coccolithophores)	2–100	106	109.5	12	4.5	1.5	0.7	86 (CPOC*)	64	76 (URPOC*)	58
Detritus (discrete, visibly, resolvable, non-living organic)	No size specification (see text)	60	36	2	1	0.5	0.3				
Total of all observed particles				62	36	9.6 (MPOC*)	5.6				

*MPOC = microscopically resolvable particulate organic carbon;
CPOC = chemically determined particulate organic carbon;
URPOC = unresolvable particulate organic carbon.

S.—B

86–88 %, is present as material which cannot be resolved by microscopy, but which is retained by glass fibre filters and is thus detectable by chemical means. This material is termed unresolvable particulate organic carbon (URPOC) (see Table 1.2).

There appears to be a semantic distinction in the literature with regard to what constitutes detritus. Reiswig (1971*b*) adopts a narrow definition which differs from that adopted by Jørgensen (1966) in his standard work on suspension feeding. Reiswig restricts the term to material which is clearly non-living, but which is still resolvable optically, while Jørgensen considers detritus to be all non-living material which is retained by glass filters. Thus detritus for the latter author includes the unresolvable particulate component enumerated separately by Reiswig.

There is only one valid reason for partitioning what is obviously a continuum of non-living particulate material which extends down in size to the colloidal range, where it probably exists in a state of physico-chemical equilibrium with dissolved material. If organisms can be shown to select within this spectrum in terms of type or size of particle that they filter, it is worthwhile to specify this component separately. This is what Reiswig attempted; he is not quite successful for, since the type of filter used sets the limit for the URPOC fraction, he cannot exclude the possibility that particles which pass through a filter may still be retained by the sponge.

(b) What do Demospongiae filter from the environment?

All three species for which we have information retain a high percentage of the resolvable particulate material which is admitted by the ostia (50 μm diameter). On average, 79 % of the calculated carbon content in this MPOC fraction is retained. If particle number in exhalant as compared to inhalant samples is taken as the index of retention, the efficiency drops to 41.9 % for all particles except bacteria, negative values indicate a net production of detrital material. It is certain that detrital material does make some contribution to diet, but the metabolism of any particles will produce detritus for excretion and this component cannot be differentiated from ambient detritus.

Tables 1.3 and 1.4 summarize the diet of three sponges and indicate percentage retentions in terms of particle size and volume.

(c) Is there any evidence of particle selection?

There is no clear evidence of selection within the spectrum of particulate material according to either shape or size. Provided that the particle has one dimension less than 50 μm, capture mechanisms operate unselectively. It is possible that armoured plankton components after ingestion are retained at lower efficiency than unarmoured cells. Reiswig argues that this is so for two species. The data are inconclusive but, if correct, no doubt reflect the resistance of cellulose and silicified walls to enzymes and the differences in time elapsed between ingestion of a given particle and excretion of its products in these three species.

Table 1.3 *Retention efficiencies of resolvable particulate organic fractions by three sponges. Percentages represent 95 % confidence intervals of the mean (data from Reiswig 1971*b*).*

Fraction	*Mean percentage retention* *Mycale* sp.	*Verongia gigantea*	*Tethya crypta*
Bacteria			
Number	96.9	94.8	96.5
Volume	97.6	95.1	96.2
Unarmoured cells			
Number	80.4	82.5	94.4
Volume	88.2	88.4	91.7
Armoured cells			
Number	41.2	38.7	66.0
Volume	64.5	77.0	88.1
Detritus			
Number	−172.3	−186.4	−148.5
Volume	−51.7	−113.5	−82.5
URPOC*	20–40	18–60 Mean 35	30–41

*URPOC=unresolvable particulate carbon.

The net production of detrital material by all species does not indicate selection against detrital size particles by capture mechanisms, but simply underlines the large output of incompletely digested cellulose walls and cell debris derived from digested plankton components.

In summary, the composition of the diet, after careful analysis, reflects the particulate composition of the water and it is clear from the high retention percentages shown for all the larger plankton fractions, that any one could surpass the URPOC as the major carbon source, if present in sufficient quantity. Arguments produced to date, on the basis of energy budgets, do not alter this fact.

1.2.3 Reproduction

The water current effects all necessary exchanges in connection with reproduction. Reproductive processes will be dealt with in detail in a later chapter, but the role of the water current in reproduction should be noted here. Some sponge species are hermaphrodite; in others the sexes are either permanently or temporally separate. However, in all cases cross-fertilization is inferred to take place although it has rarely been observed. Sperm are released into the exhalant stream

Table 1.4 *Net diet of three sponges in terms of particulate organic carbon removed from each particulate fraction per unit of sea water filtered (data from Reiswig 1971b).*

Fraction	*Carbon source in mg m*$^{-3}$ *Mycale* sp.	*Verongia gigantea*	*Tethya crypta*	*% total retention mean of 3 species* *MPOC*	*POC*
Bacteria	0.17	0.14	0.55	4.6	0.9
Unarmoured cells	3.77	4.18	6.07	83	16.2
Armoured cells	0.38	0.62	1.19	12.3	2.4
Miscellaneous (filamentous blue-green algae and fungi)	negligible	negligible	0.01		
Detritus	−0.04	−0.12	−0.38		
MPOC*	4.29	4.82	7.43		19.5
URPOC*	15.81	21.28	32.28		80.5
Total POC*	20.1	26.1	39.7		100.0

*MPOC = microscopically resolvable particulate organic carbon; URPOC = unresolvable particulate organic carbon; POC = particulate organic carbon.

and taken in with the inhalant stream of another individual, trapped by choanocytes and transferred to the eggs which lie in the mesohyl. In species which are oviparous, eggs are extruded either by way of the exhalant current or directly by dissolution of the dermal membrane. In viviparous species, larvae or tiny adult sponges are similarly expelled in the exhalant stream.

1.2.4 Respiration

Respiratory exchanges between the water current and the cellular sites where oxygen is utilized take place by diffusion. In sponges there is a large area of naked cell surface available for exchange along all inhalant canals, over the general body surface, and at collar tentacles and cellular surfaces of choanocytes. Diffusion is an effective method of oxygenation when the diffusion path is less than 1.0 mm; in no case in a sponge would the diffusion path exceed this distance.

Little unanimity is to be found in the literature with respect to respiratory rates in Porifera. This is not surprising since all the standard factors which increase or decrease respiratory rates operate in sponges indirectly through their effect on pumping activity. As we have noted, this is invariably lowered in laboratory experiments sometimes by an order of magnitude (Table 1.5). This source of error plagues laboratory studies on all aspects of sponge activity, but has been compounded in experiments which have attempted to determine respiratory rates by the use of closed respirometers. These are unsuitable devices for estimation of sponge respiration, since the depressive effects of water recycling and exposure to waste cannot be avoided and the respiratory activity of attached surface organisms is incorporated in the results. Attempts to mitigate the effects of unnatural conditions by using very small sponges suffer from increased error from the latter source.

With the exception of the recent study by Reiswig (1974) where oxygen utilization in relation to body size and pumping activity was estimated from in-situ measurements, only Hazelhoff (1938) has estimated respiratory exchanges by monitoring oxygen differences between inhalant and exhalant water. Hazelhoff, however, worked with laboratory animals and did not attempt to determine rates of water pumping, thus his results are difficult to compare with others.

In Table 1.5, a comparison is made of data on respiration in Porifera. Of the older studies using respirometers only those on Calcarea yield results comparable to those obtained in field populations of Demospongiae. In all experimental situations the smaller, more simply constructed Calcarea are affected less than the Demospongiae. A general depression of oxygen utilization is shown in respirometer experiments, while Hazelhoff's analysis of water samples shows a consistently higher rate of oxygen removal in all species studied. Reduced respiration rates indicated in the respirometer studies result from a decrease in water pumping activity, which is naturally accompanied by an increase in the percentage of available oxygen removed from the water that is being circulated.

In slow moving and sedentary animals, oxygen consumption and utilization

Table 1.5 *Respiratory rates in Porifera.*

Species	*Reference*	*Oxygen utilization in ml* O_2 h^{-1} *per unit wet wt in g*	*per ml wet volume*	*% available* O_2 *removed*
Calcarea				
Grantia sp.	Hyman 1925	0.10–0.15		
Sycon raphanus	Hazelhoff 1938			39
Leucandra aspersa	Hazelhoff 1938			11
Demospongiae				
Chondrosia reniformis	Hazelhoff 1938			21
Tethya aurantium	Putter 1914	0.0074		
Suberites domuncula	Hazelhoff 1938			15
Suberites domuncula	Putter 1914	0.0078		
Haliclona rubens	De Laubenfels 1932		0.55 (light)–0.15 (dark)	
Haliclona rosea	Hazelhoff 1938			6
Iotrochota birotulata	De Laubenfels 1932		0.067 (light)–0.154 (dark)	
Ephydatia fluviatilis	Kilian 1964		0.043	
Verongia aerophoba	Hazelhoff 1938			57
Hippospongia equina	Putter 1914	0.0053		
Mycale sp.	Reiswig 1974	0.126	0.0444	1.17
Verongia gigantea	Reiswig 1974	0.1004	0.0628	5.6
Tethya crypta	Reiswig 1974	0.0329	0.0154	0.89

is always low, no more than 20 % oxygen withdrawal from the respiratory current has ever been demonstrated reliably in sponges, lamellibranchs or tunicates (Nicol 1960). The higher percentages recorded by Hazelhoff for some sponges are certainly artefacts caused by reduction in pumping rate. Reiswig's results for sponges accord well with this general, logical tendency and are at this time the only observations on respiration rates in Demospongiae which need no qualification.

1.2.5 Activity

It has been assumed that sponges, with their simple unidirectional water current driven by the unco-ordinated beating of the choanocyte flagella, sustain a constant level of pumping activity under normal conditions (Jørgensen 1966). On the other hand sponges can respond to adverse or experimental stimuli by closure of ostia, canal constriction, backflow and reconstruction and compression of flagellated chambers. The effect in all cases is to immobilize the choanocyte flagella. This type of response may simply be passive as in a case where the amount of suspended material in the water is sufficient to occlude the inhalant system. It could, however, in response to less extreme conditions, be a controlled or truly endogenous response, and in either case be mediated by cell-to-cell transfer of contractile stimuli. If so, such modulations in activity may occur during normal pumping cycles.

Naturally, studies concerned with the rate of filtering in sponges have attempted to relate activity of the choanocytes to particle concentration, and unless the latter becomes physically limiting, when the sponge will contract and immobilize the choanocyte flagella, then filtering rate does appear to be independent of particle concentration. This does not mean that activity is always sustained at a constant level, for requirements other than filtering must be met by the water current. For example, as growth proceeds and the established surface to volume ratios alter, hydrodynamic and metabolic factors will require canal reorganization. During this process, activity level decreases. Periods of reproductive activity cause substantial drops in the level of pumping activity as many choanocytes are expended in gamete production. Quite apart from these situations where processes other than feeding exert an influence on the pumping activity, there is evidence that under normal conditions variations in pumping activity and flow rate occur.

First, it has been recognized for a long time that the choanocyte chambers of many Demospongiae and some Calcarea contain a specialized 'central' cell. The role of these cells was not understood until ultrastructural studies were completed by Connes, Diaz and Paris (1971), who showed that the surfaces of central cells had deep grooves in which choanocyte flagella became trapped. As the central cell moved within the choanocyte chamber, from central to apopylar position, it immobilized flagella, forced part of the water stream to pass through its surface grooves and reduced the effective lumen of the apopyle (Fig. 2.2, p. 59). Since most capture of suspended material takes place at the choanocyte surface, the intervention of a central cell to retard current flow through the choanocyte chamber allows more time for ingestion to occur. This is an example of structural specialization which allows the modulation of flagellar activity, and hence pumping activity, by simple mechanical means.

Secondly, studies in the field where exhalant current velocity and oscular diameter were monitored and related to obvious environmental variables such as temperature, wave action, turbidity and sedimentation have been carried out on three species of Demospongiae (Reiswig 1971*a*). Considerable variation exists in the observed patterns of activity. *Mycale*, a small, thin-walled, tubular sponge sustained a constant level of water transport over the successive experimental periods of 21–75 hours. *Verongia gigantea* is a large, thick-walled species with mesohyl matrix reinforced strongly by fibrillar collagen; over the experimental period of 23 to 141 hours *Verongia* maintained a high level of pumping activity for long intervals and then interpolated brief periods, averaging 42 min., of complete cessation. This behaviour cannot be related to any environmental factors and the morphological site where cessation is initiated has not been identified. A specimen 660 ml in volume can pass from normal activity to complete cessation in 3.8 to 6.4 min.; this appears to be an entirely endogenous behaviour pattern. The third species, *Tethya crypta*, is a solid spherical sponge inhabiting water shallower than that in which the previous species is found. It displays a very complex activity pattern which has two major components, a diurnal activity/inactivity cycle superimposed on which are periods of complete contraction occurring at intervals of 9–21 days and lasting for 1–5 days. As in the case of *V. gigantea*, variations in the pumping behaviour of *T. crypta* do not correlate with environmental variables but appear to be endogenous.

The implication of these observations with respect to transmission of stimuli in sponges will be considered in relation to sponge histology, but at this juncture they serve to illustrate the considerable variation in patterns of sponge pumping activity and the ability to interrupt pumping in response to endogenous stimuli. Variations in pumping rates of sponges and the relationship of pumping rate to oxygen utilization are presented in Tables 1.6 and 1.7. In Table 1.6, the figures for pumping efficiency, expressed as litres of water pumped per millilitre of oxygen consumed, were derived from field observations, and in the case of *T. crypta* and *Mycale* sp. compare very well with values calculated by Jørgensen (1966) for *Halichondria panicea* and *Sycon compressa* on the basis of dispersed literature reports. *Verongia gigantea* is far less efficient; it is, however, different from other species so far studied in that it supports a concentrated population of bacterial symbionts, and the contribution these make to total exchanges is not easily determined.

Comparing sponges with other filter-feeding organisms, *Mycale* sp. and *T. crypta* are more efficient than bivalves, tunicates and bryozoa in terms of litres of water pumped per millilitre oxygen consumed.

It is becoming apparent that sponges are not as simple as is customarily supposed, but are complex organisms which can respond directly to environ-

Table 1.6 *Pumping rates of sponges.**

Species and reference	*Pumping rates expressed in ml water pumped per second, per:* *ml sponge (volume)*	*g wet wt (fresh wt)*	*g dry wt*	*g ash free wt (organic matter)*
Mycale sp. (Reiswig 1971*a*)	0.266	0.648	3.47	5.01
Verongia gigantea (Reiswig 1971*a*)	0.087	0.1225	0.766	1.165
Tethya crypta (Reiswig 1971*a*)	0.154	0.264	1.03	2.76
Leucandra aspersa (Bidder 1923)	0.0368		0.44	1.24
Grantia compressa (Jørgensen 1955)		0.178		0.165
Halichondria panicea (Jørgensen 1955)		0.103		1.33
Suberites domuncula (Putter 1914)	0.0023	0.0023	0.016	0.0324
Ephydatia fluviatilis (Kilian 1952)	0.28			

*Four different ways used by Reiswig (1971*b*) to express his data allow comparison with earlier less complete work.

mental and endogenous stimuli and are capable of a low level of co-ordinated activity. It remains to be demonstrated that this co-ordination is the product of an elementary integrative system, there are no observations of co-ordinated activity occurring with sufficient speed to necessitate neuronal activity as it is understood in higher organisms.

1.3 The form of sponges in relation to the water current

The water current serves for food gathering and tissue oxygenation and, at the same time, to expel excretory and reproductive products, thus the optimum morphology for any sponge will be that which ensures maximum separation of the incoming surface water from the exhalant stream leaving the oscules. Bidder (1923) noted that in terms of pressure and velocity of the water current, there will be, for any given sponge, an optimal oscular or atrial diameter; it is

Table 1.7 *Pumping efficiency in relation to respiration.*

Species	*Reference*	*Rate of water pumping*	*Oxygen utilization*	*Efficiency of pumping*	*Comments*
Verongia gigantea	Reiswig 1971*a*	Own volume in 9 s. 0.087 ml /ml of sponge /sec.	Consumes 5.6% of available O_2. 0.196 ml O_2 $kcal^{-1}$ as tissue h^{-1}	4.08 litres ml^{1-1}/O_2	All data in-situ measurements on population
Mycale sp.	Reiswig 1971*a*	Own volume in 3.8 s. 0.266 ml /ml of sponge /sec.	Consumes 1 % available O_2. 0.432 ml O_2 $kcal^{-1}$ as tissue h^{-1}	19.57 litres ml^{-1}/O_2	All data in-situ measurements on population
Tethya crypta	Reiswig 1971*a*	Own volume in 6.5 s. 0.154 ml /ml sponge /sec.	Consumes 1 % available O_2. 0.113 ml O_2 $kcal^{-1}$ as tissue h^{-1}	22.78 litres ml^{-1}/O_2	All data in-situ measurements on population

that which ensures that the exhalant jet travels as far as possible from the inhalant surface. The physical factors controlling the exhalant velocity are the diameter of successive pore and canal components, the pressure generated in each chamber by the choanocyte flagella, and the frictional resistance of the canal walls. Bidder's hypothesis has been substantiated by observations on change in oscular diameter in relation to velocity of the exhalant stream in *Tethya crypta*. In this sponge, no matter how much the volume of water flowing through the system varies, oscular contractions maintain the exhalant velocity within narrow limits and thus ensure effective removal of exhalant water.

At the level of overall body form, several strategies which serve to maximize separation of the water streams are apparent. Many Demospongiae are elevated on a stalk and can then be oriented to have an ostial and an oscular face or side; many are cup shaped, with convex lower inhalant and concave upper exhalant surfaces (Pl. 8*c*). In these cases separation of water streams is achieved either by orientation to a current or by deflection by the rim of the cup. Small specimens of *Microciona* reconstituted from cell suspensions in still water organize an apical oscular tube; if reconstituted in a current, the oscular tube bends to discharge with the current.

Generally, oscules tend to be apical and oscular velocities high – this suffices for sponges in turbulent shallow waters where no current direction is sufficiently predictable to make stable morphology advantageous. In deep-water environments, many species of both Demospongiae and Hexactinellida exhibit stable, externally symmetrical body form in response to the stability of the environment (Figs. 1.5, 1.7).

1.4 Evolutionary strategies in different lines of Demospongiae: the major morphological developments

Increase in volume in all sponges is achieved by increase in volume of the mesohyl and elaboration of its skeletal content. The choanoderm and pinacoderm remain one cell deep, and by continued folding maintain a surface to volume ratio sufficient to sustain respiratory exchange throughout the whole structure.

In the simplest Demospongiae, the choanoderm is leuconoid, subdivided into chambers which are aggregated in groups; the pinacoderm is continuous at the surface; and between the two layers the thin mesohyl contains a dispersed collagenous skeleton and supports a spicule skeleton of both megasclere and microsclere components. In such a sponge, for example *Plakina* (Homosclerophorida) (Fig. 1.7*a*), the amoeboid cell components of the mesohyl are few in number and uniform in structure. In a sponge such as this which is always a thin incrustation, mesohyl thickening can occur in three positions; basally, between aggregates of choanocyte chambers, or superficially. *Plakortis* (Fig. 1.7*b*), a close relative of *Plakina*, demonstrates mesohyl thickening accompanied by cellular diversification. The cortex formed as a result of superficial mesohyl thickening can be reinforced by concentrations of fibrillar collagen as in *Plakortis*, or can serve

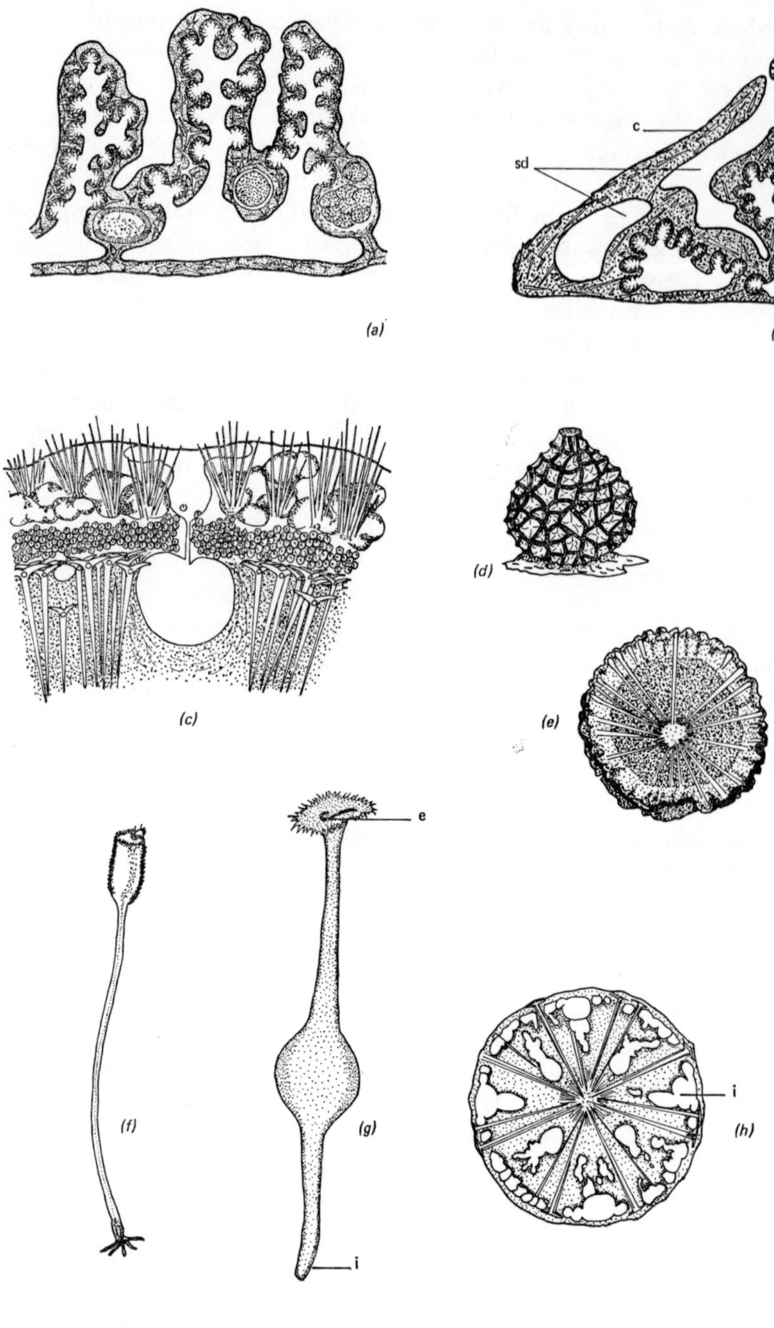

(a)
c
sd
(b)
(c)
(d)
(e)
e
(f)
(g)
i
i
(h)

Fig. 1.7 The body form and organization of some Demospongiae.
(*a*) A section through *Plakina* (Homosclerophorida) at right angles to the surface. This sponge has a very simple organization. The body is greatly folded but the mesohyl remains thin. Embryos are shown toward the base of the sponge, a position they always occupy in encrusting forms.
(Redrawn from Schulze 1880.)
(*b*) A section through *Plakortis* (Homosclerophorida) at right angles to the surface. In this sponge, the mesohyl has become thicker superficially, and the choanocyte chambers are separated from the surface by a cortex (c) and subdermal lacunae (sd).
(Redrawn from Schulze 1880.)
(*c*) A section through *Geodia* (Choristida) at right angles to the surface. In this group the mesohyl is large and complex. Spicule elements make up a large proportion of the sponge weight, and are disposed in a radial fashion. Special cortical spicules provide a superficial armour.
(Redrawn from Sollas 1888.)
(*d*) A specimen of *Tethya* showing the characteristic globular body and regular inhalant pore grooves over the surface.
(*e*) *Tethya* in transverse section to show the strict radial disposition of the spicule skeleton and the marked difference between the superficial cortex and the deeper endosome. The centre of the sponge where the spicule tracts converge is solid spicule.
(*f*) *Stylocordyla australis*, a deep-water member of the Demospongiae with symmetrical body form and internal radial spicule arrangement.
(*g*) *Disyringia dissimilis*, a deep-water member of the Demospongiae which has a very symmetrical body with specialized inhalant (i) and exhalant (e) poles.
(Redrawn from Sollas 1888.)
(*h*) A transverse section through the central body region of *Disyringia dissimilis*. The radial disposition of the spicules is shown. The separation of superficial (inhalant) canals (i) and deeper (exhalant) canals can also be seen.
(Redrawn from Sollas 1888.)

as a location for a solid spicule skeleton which in some cases becomes an extremely dense protective structure. Both types of reinforcing permit the support of a thicker sponge body, but in the group exemplified here the overall habit of the sponge remains spreading to subspherical rather than becoming erect or foliose.

If now the components of the spicule skeleton are diversified and organized, both regionally and in terms of orientation, the result is to emphasize the cortex as distinct from the choanosome. The cortex is usually packed with microscleres or debris superficially and contains tracts of megascleres which are oriented in a radial manner with respect to the central basal point of the sponge. Radial construction can become extremely pronounced and characterizes all regions of the sponge, as in *Tethya* (Figs 1.7*d*, *e*) and *Cinachyra*, or it may be recognizable only in the cortex and outer regions of the choanosome, as in *Geodia* (Choristida) (Fig. 1.7*c*).

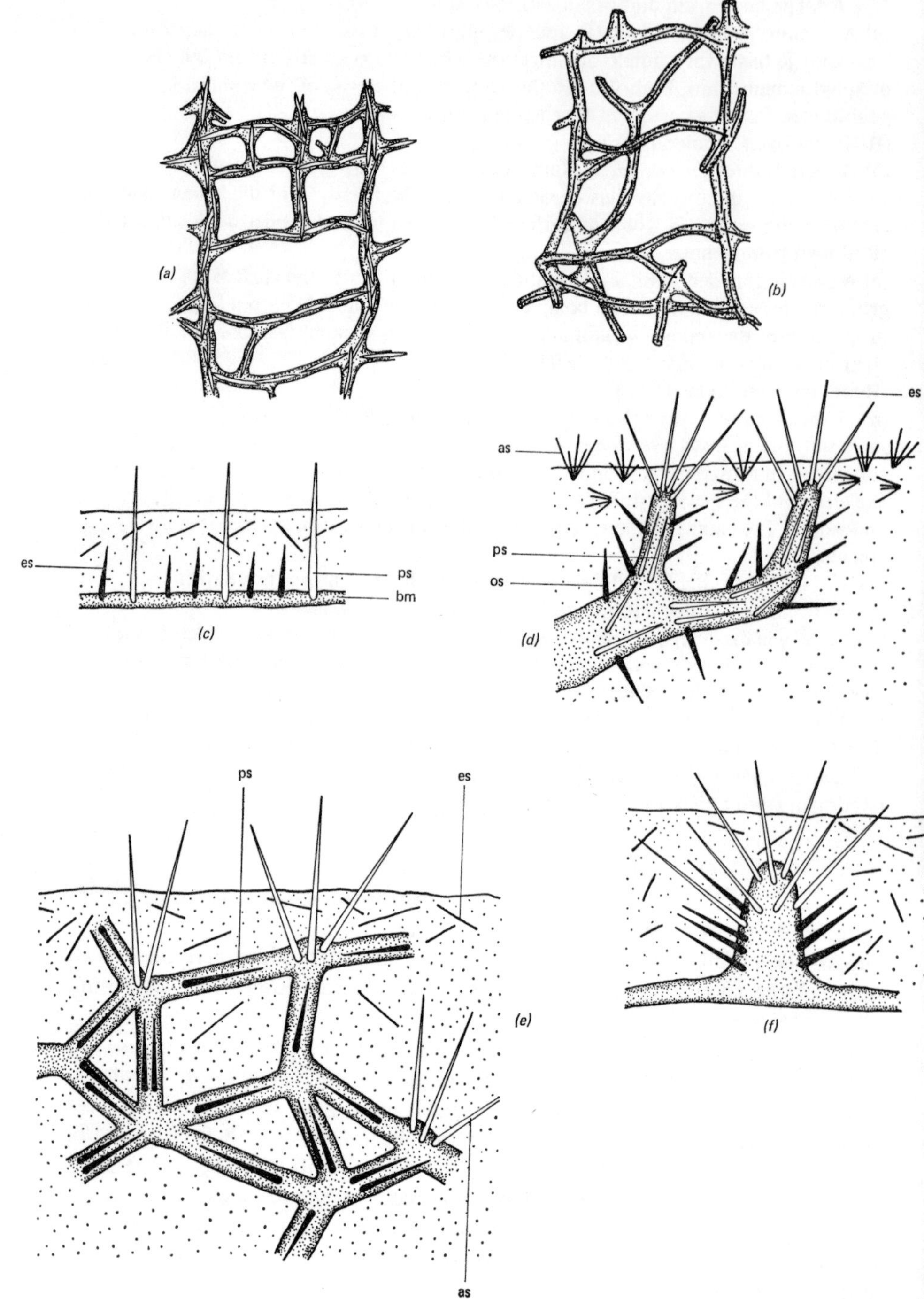
(a)
(b)
es
ps
bm
(c)
es
as
ps
os
(d)
ps
es
(e)
as
(f)

Fig. 1.8 Some examples of skeletal organization in Demospongiae which elaborate spicule and fibre skeletons (Dark spicules are spiny; clear spicules are smooth.)
(*a*) A section at right angles to the surface of *Haliclona* to show a regular spongin fibre network with some spicule reinforcement. This skeletal arrangement is also typical of the family Callyspongiidae.
(*b*) A section at right angles to the surface of *Chalinula* to show the irregular spongin fibre network which contains very few spicules.
(Redrawn from Gressinger 1971.)
(*c*) A diagrammatic representation of the skeletal arrangement in some encrusting Poecilosclerida where a basal spongin fibre mat (bm) anchors the sponge to the substrate and supports the primary spicules (ps) and the smaller echinating spicules (os). Such a skeletal arrangement is found in the genus *Microciona*.
(Redrawn from Levi 1960.)
(*d*) A diagram of a plumose spongin fibre skeleton which has large fibres fanning out toward the sponge surface. Four distinct spicule types are differentiated here: those at the surface, the accessory spicules (as); those embedded completely in the fibre (ps); those terminating the fibre and piercing the surface, the ectosomal spicules (es); and those embedded in, but protruding from, the fibre, the echinating spicules (os). This precise type of skeleton occurs in the genus *Rhaphidophlus* and with minor modifications typifies all erect, branching genera of the family Clathriidae.
(Redrawn from Levi 1960.)
(*e*) An isodictyal spongin fibre skeleton with triangular fibre mesh. The spiny spicules (dark colour) are now incorporated in the fibre and are the primary spicules (ps), tufts of accessory spicules (as) occur at the intersections of fibres near the surface and the ectosomal spicules (es) are strewn without order at the surface.
(Redrawn from Levi 1960.)
(*f*) Skeleton of an encrusting sponge which shows some elaboration of the basal mat seen in (*c*). Fibres are beginning to develop and carry the same types of spicules seen in the basal plate. This development is intermediate between types (*c*) and (*d*). Isodictyal development is quite separate. (Redrawn from Levi 1960.)

Forms with pronounced radial megasclere orientation are frequently spherical with small base of attachment and are almost solid spicule at the centre. Those like *Geodia* where this radial construction is more superficial can be spherical, but frequently are irregular ramifying sponges which attain great size. The spicule cortex is often reinforced by the deposition of fibrillar collagen in which contractile cells aggregate, this cortex is sufficiently contractile to control the diameter of inhalant and exhalant apertures despite the presence of a heavy spicule armour.

In sponges with a radial organization such as *Tethya*, where architecture is very dense, it is usual to find only one or two apical exhalant apertures and to find ostia organized into special ostial or sieve areas. In *Tethya*, ostia are sunk in grooves roofed by a thin pinacoderm and stretched between the spicule and collagen columns of the cortex. In some sponges, very elaborate and stable inhalant structures are developed; these are not restricted to forms with a highly organized cortex but are most common in such groups. The most extreme example is that of *Disyringia dissimilis* (Figs 1.7*g*, *h*) where inhalant and exhalant tubes are differentiated at opposite poles of a radially constructed, spherical body. Development of stable inhalant cribripores as in *Polymastia* (Pl. 9*b*), or porocalyces as in *Cinachyra*, represent the closest approach to organ formation to be found in the Porifera.

A quite different solution to organizing support for a larger body volume is seen in those sponges which develop a proteinaceous fibre skeleton in conjunction with a spicule skeleton. The latter can, to varying extents, be enclosed within the collagenous fibre which is known as spongin. Different permutations of this highly successful type of organization will be mentioned in the course of explaining classification but, in general, this skeletal structure produces a sponge with open architecture with a large volume of canal space and low density of mesohyl matrix, the latter always lightly reinforced by fibrillar collagen. The form of these sponges is frequently branching or tubular, but no general rule applies; spicules can be diversified and localized, but there is no radial arrangement. Examples of this type are *Callyspongia* (Fig. 1.8*a*; Pl. 5*b*) where the fibre predominates and *Clathria* where spicules are diverse and fibre still very abundant (Fig. 1.8*d*).

Another major group of sponges develops no mineral skeleton in the mesohyl, the sponge relies for support upon the spongin fibre skeleton and on a mesohyl matrix which may be reinforced by dense aggregations of fibrillar collagen. The architecture of such forms ranges from very diffuse as in *Aplysilla* to very compact as in *Pseudoceratina*, depending upon the density of fibrillar collagen deposition and on the degree of anastomosis of the spongin skeleton. There is no cortex in these sponges. The superficial regions are supported by identical material to that which supports the choanosome; the only distinction which marks a cortex is the density of fibrillar collagen and the aggregation of certain cell types. The term 'ectosome' is sometimes reserved for superficial regions without a special skeletal support.

From this brief outline it is clear that in Demospongiae increase in complexity of the mesohyl has proceeded in several quite different directions, and to a great extent the type of organization adopted provides a reliable guide to phylogenetic relationships within the group. However, increase in size which can accompany this elaboration is never a result of the repetition of modular units. The end product of growth in all cases is a complex sponge with no regions defined except in relation to the whole. External appearances of symmetry, where they exist, as for example in *Tethya* and *Stylocordyla* (Figs 1.7 *d*,*e*,*f*), do not extend to the aquiferous system or to the skeleton.

The Hexactinellida in their predominantly deep-water environment often achieve symmetrical external form, and simple Calcarea appear to be radially symmetrical. Larger more complex Calcarea can comprise these simple units, which give every appearance of being independent except at the point of junction with a basal attachment or branching stolon (Pl. 8*b*). Such complex forms of *Leucosolenia* (Fig. 1.3*a*) have often been cited by authors concerned to establish the colonial nature of sponges.

It is true that Hexactinellida and Calcarea are far more regular in structure than Demospongiae and that simple Calcarea exhibit an approach to radial symmetry even in the arrangement of the choanosome. In a structure where all cells are free to migrate this symmetry is not, and cannot be, absolute and hence cannot be used to support an argument for the existence of structural modules. True radial symmetry in sponges has been demonstrated to occur only in developmental stages of some Calcarea where the '*cellules en croix*' have a tetraradial disposition.

1.5 Individuality: what constitutes a sponge individual?

After their animal nature was established beyond doubt in the early nineteenth century, much of the interest in sponges centred around debating the limits of a sponge individual, despite an abysmal lack of knowledge of their structure and function.

To the present time this continues as a sporadic debate which still lacks many facts and makes many assumptions. There have been four major theories which have enjoyed popularity at one time or another. Two, which have been abandoned as serious alternatives for a long time, held either that an individual was synonymous with a single cell or with a choanocyte chamber. The first theory would make all sponges colonies and, according to the second, all but the simplest Calcarea would be colonies.

The two remaining theories have been more persistent. One suggests that since the result of larval metamorphosis is the production of a small sponge with a single osculum, regardless of the disposition of the flagellated surfaces, this osculum is the organizational centre of the sponge. In functional terms this means that the osculum, the excurrent canals draining to it, and the sector of the inhalant surface and canal system which feeds the related choanocyte chambers constitute an individual. Increase in size in sponges leads almost invariably to increase in number of oscules, a specimen of *Haliclona*, for example, with fifteen oscules would be interpreted as a colony composed of fifteen individuals.

This theory, suggested first by Schmidt (1864), was developed in some detail by Minchin (1900) in his contribution to Lankester's *A Treatise of Zoology* and adopted by Hyman (1940) although the latter author lamented the lack of facts which would help in deciding between theories. It is not surprising that the idea that 'a sponge individual is represented by the osculum and its contributing parts' has become the most frequent interpretation of sponge individuality in the English language literature, since the works by Minchin and Hyman are the only comprehensive accounts of sponge structure and biology in this language up to the present time. Recent authors who adopt this view are Hadzi (1966) and Brien (1967, 1973).

If oscules are held to be the external indicators of physiological units in a colonial structure, they can only be so regarded by homologizing each unit with a simple, unbranched, asconoid calcareous sponge. However, as sponges grow new oscules, inhalant and exhalant systems arise in response to the hydrodynamic pressures in the sponge as a whole; they do not arise by a process of budding or by outgrowth and fission of pre-existing systems. The origin of a new osculum is a growth process, not in any way to be compared with budding, which is an asexual reproductive process. There is no evidence of a relationship between extant and newly formed oscules in terms of tissue anlagen, in the same sense that a bud of a cnidarian carries an extension of the parent gastrovascular cavity.

The evidence that we have on the location of new oscules in relation to existing units suggests that they arise marginally, the sponge grows normally by extending the area of a cell layer of more or less constant thickness and new systems arise in the area of extension. In forms with a stable massive, rather than tubular, encrusting or ramose shape, tissue thickness may increase without marked extension of surface area or increase in number of surface canals and oscules. Here, existing canals simply increase in diameter to handle larger volumes.

The fourth interpretation of the individual in sponges is that which holds that the individual is all the substance bounded by a continuous pinacoderm. This is not a recent view, the naturalist Dujardin held very similar ideas, and recently Tuzet, Pavans de Ceccatty and Paris (1963), Borojevic (1968) and Hartman and Reiswig (1973) have discussed and upheld this interpretation.

Many minor points could be considered, but the most significant arguments in favour of the isolated sponge being the individual stem from our increased understanding of the functional morphology and behaviour of sponges. It is now quite untenable to consider sponges as colonies where oscular modules are independent of each other. This view could be only imperfectly upheld on the basis of static morphology of the simple Calcarea; it cannot suffice if we recognize that a branch (equivalent to an oscular unit) of *Leucosolenia* shares common mesenchymal elements with all other branches, that within the whole structure elements move freely, particularly reproductive elements, that the sponge can remodel and reroute its aquiferous system and, finally, that it can fuse with neighbouring sponges. Behaviour of this type will permit only a dynamic interpretation of an individual, the definition must accommodate the reciprocal relationships of all the cells confined at any particular time within a bounding pinacoderm.

If a sponge was a colony, a polymer of oscular units, these units should behave independently of one another with respect to phases of contraction and expansion and to water flow. In encrusting sponges, where exhalant canals travel to a great extent at the surface (Pl. 5*a*), it is easy to observe directly that adjacent oscular systems are not independent; also, if one oscule closes while others remain open, water currents simply bypass the closed aperture and flow through the open one. In massive sponges with loose architecture the typical morphology is a system of intercommunicating subdermal lacunae, and many oscular systems coalesce at this level (e.g. *Halichondria, Hippospongia*).

The existence of co-ordinated behaviour in sponges has been demonstrated with respect to synchronous oscular contraction and dilation in *Verongia gigantea* and *Tethya crypta*. Rhythms of oscular expansion and contraction do operate, and the entire population of oscules in a sponge behaves in concert (Reiswig 1971*a*). Similarly melanization of cells in *Leucosolenia* and *Clathrina* in response to either endogenous or artificial stimuli occurs throughout the whole specimen; component tubes do not behave independently (Pavans de Ceccatty 1958).

If external stimuli were implicated in the above responses, it could be argued that the same stimuli were acting on all oscular units. However, in the case of oscular rhythms no environmental effects were involved, and the behaviour was endogenous, an attribute of the whole sponge, an attribute of an individual.

It is important at this juncture to emphasize again the over-riding importance of the water current in sponges and to realize that the number of exhalant apertures in any sponge is a function solely of optimization of the water flow, that new oscular systems arise as growth proceeds and as the surface to volume relationships in the mass change. Addition of new oscular openings and canals is not polymerization or a budding process, it is a growth process. Therefore the number of oscules is only a pattern which has no long-term stability, since it is dictated at all times by a number of factors which are constantly changing. Certain physical factors must be optimized, flow rate maintained, canal length kept to a minimum to reduce frictional effects and separation of inhalant and exhalant streams ensured; this has all to be achieved in a mass where no cells are immobile.

Time-lapse cinematographic observation of fresh-water sponges provides graphic evidence of the flexibility of canal systems and the disposition of oscules. Also, in fresh-water sponges, it has proved possible to excise oscular tubes, place the resultant sphere of tissue on an area of pinacoderm and within a few days induce the formation of an osculum at the contact site. It is not yet clear what this demonstration of induction means in terms of differentiation, but it serves to dispel any notion that an osculum is an extension of and represents an underlying unit individual.

So far the concept of a sponge as a colony has been discussed and rejected mainly with reference to sponge structure and function. Compare for a moment the type of structure found in the cnidarian polyp, where there is an undoubted colonial organization, with that seen in Porifera. Cnidarian polyps show regional specialization within the tissue layers, buds carry extensions of the parent gastrovascular cavity, there is a permanent relationship between apertures, and internal symmetrical structures such as septa are common. No structures analogous with these occur in Porifera, where there are no recognizable subunits. Individuality in sponges is recognizable only at the level of entire specimens invested by a continuous pinacoderm.

2 Sponge cells: their structure and behaviour

2.1 Introduction

All authors concerned to establish a system of description and nomenclature which can be applied to sponge cells must despair over the confusion which exists in sponge histology.

This is a field in which great progress has been made recently, as a result of ultrastructural and histochemical studies; but the same techniques are revealing more and more cellular diversity, making it increasingly difficult to present a simple but comprehensive synopsis of cell types. At present there is only a limited measure of agreement to be found among specialists with regard to the nomenclature and functions to be ascribed to particular combinations of cellular characteristics. The characteristics themselves can, however, be described precisely. There are several reasons for the persistent problems, but most important is the fact that sponges have no permanent histological system; they have, at all times, a functional system within which probably all components are mobile, and many can change structure. This temporal instability in terms of both position and structure makes it difficult to study particular cells over any time period. With the exception of cells which become filled with very characteristic inclusions, the only convincing evidence of the sequences of differentiation of which given cells are capable comes from work on the development of larvae and gemmules, the formation of gametes, and from experiments on the reconstitution of sponges from dissociated cell suspensions. In these circumstances there is some measure of predictability and control, in that the sponge is undergoing a directional differentiation leading to the formation of a recognizable end product, either an embryo or a small sponge. The adult sponge has no such characteristics.

What most needs to be determined at present is the extent to which sponge cells remain multipotent, despite at any given time appearing to have become specialized for particular activities.

It is futile in the long term to attach names to cell states, the inter-relationships of which are not understood; however, in the short term it is very desirable that we be able to describe the histology of a sponge, and know that other workers will concur in the nomenclature and function ascribed to particular cells.

For this reason the structure and functions ascribed to the major cell types of adult sponges will be detailed. The transformations which are known to take place in the course of differentiation, and the behaviour of sponge cells in aggregation can then be considered.

2.2 Cell types in Demospongiae

It has been possible to confirm the function of most of the cell types of sponges only since ultrastructural information has become available; thus we shall be considering, almost exclusively, ultrastructural morphology. In the absence of recent studies it is impossible to say anything about the Hexactinellida. The Calcarea also have been very little studied at the ultrastructural level; they compare more closely to the Demospongiae, but show less cellular diversity.

There are four ways of ascribing particular functions to sponge cells:

(i) Position: if this is constant it should provide some clue to function.
(ii) Structure.
(iii) Determination of the chemical nature of cellular inclusions.
(iv) Observations of cell behaviour.

We can now comment on the first two attributes for most cells, and, although facts on histochemistry and behaviour are still scarce, where available they are useful.

2.2.1 Cells which line surfaces (Table 2.1)

(a) Pinacocytes

The pinacoderm is a continuous layer which, when viewed from above, has the appearance of a squamous epithelium (Pls. 12*b*, *d*). It separates the sponge mesohyl from the environment, lines the external surface, forms the basal attachment lamina and also lines all exhalant and inhalant canals. The pinacocytes which constitute this layer show some structural and functional differences which coincide with their location, either at the surface (exopinacocytes), lining canals (endopinacocytes), or basal (basopinacocytes).

Exopinacocytes (Figs 2.1*a*, *b*)

In the great majority of sponges these cells have a 'T' shape with the flattened extensions in the plane of the surface and with the cell body, containing the nucleus and most of the vacuoles and inclusions, pendant into the mesohyl. In Calcarea, typical 'T' pinacocytes are accompanied by fusiform types, where the extension is in the plane of the surface and the nucleus is centrally placed in the same plane. Some Demospongiae appear to have only this fusiform type of pinacocyte. Exopinacocytes are well supplied with mitochondria and lysosomes; they contain phagosomes and triglyceride-positive vacuoles, and have a well-marked Golgi apparatus and cytoplasmic microfilaments. The nucleolus is present or absent unpredictably. Since the endopinacoderm is the bounding layer of a sponge, much effort has been directed towards identifying in this layer attachment complexes between adjacent cells – structures analogous either to septate desmosomes, terminal bars or tight junctions of other tissues. So far,

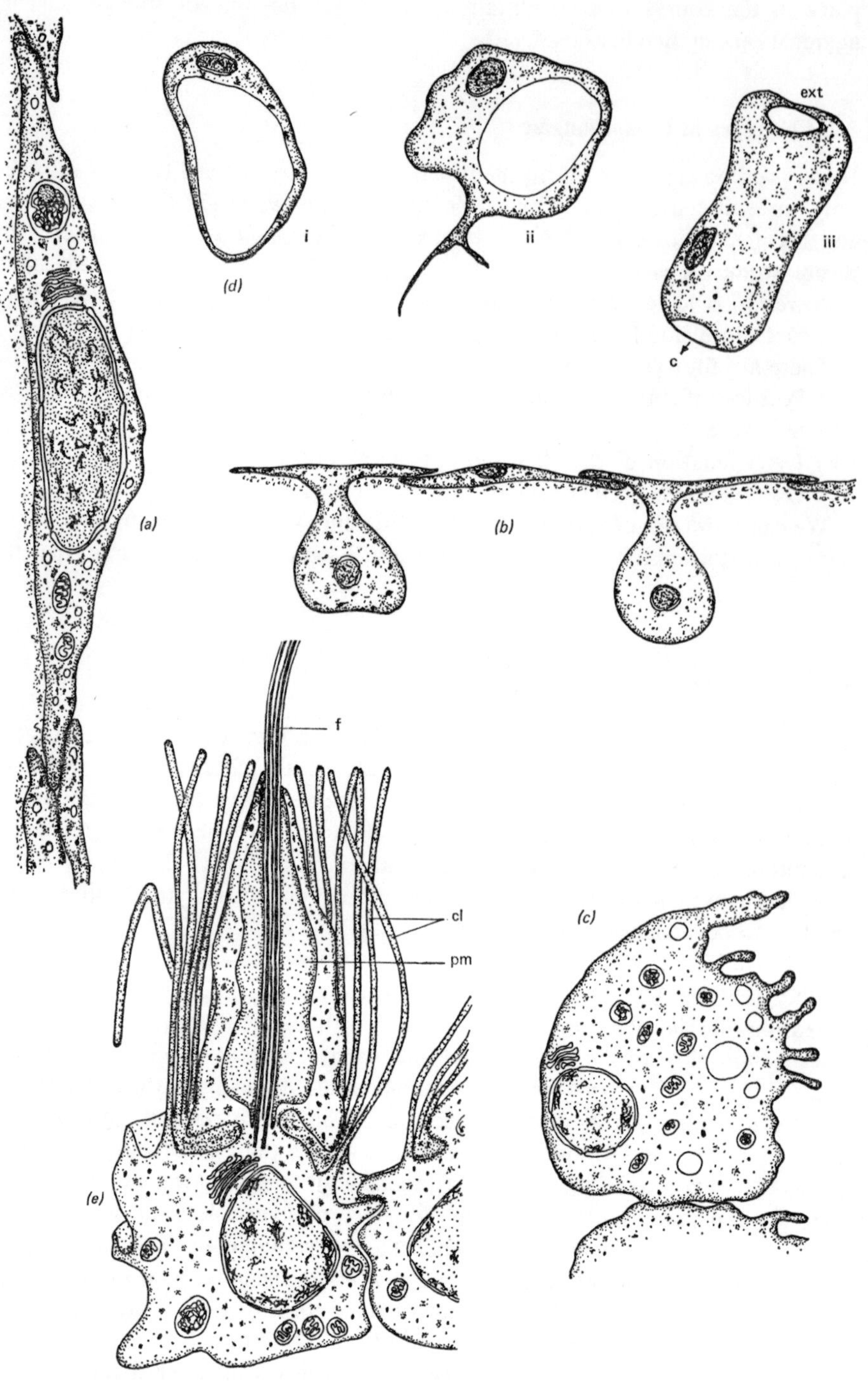
(a)
(b)
(c)
(d)
i
ii
iii
ext
c
f
cl
pm
(e)

Fig. 2.1 Cells of epithelial surfaces.
(*a*) An exopinacocyte of *Halisarca* drawn from an electron micrograph. The outer surface is covered by the polysaccharide-rich cell coat. The cell is fusiform with oblique areas of overlap between adjacent pinacocytes.
(*b*) Exopinacoderm of a calcareous sponge which has an alternating arrangement of 'T' and fusiform pinacocytes.
(*c*) Endopinacocyte of *Hippospongia*. The outer surface which faces the water canal is smooth; the inner, matrix surface is irregular. The area of contact between adjacent cells is less organized and less extensive than similar areas between exopinacocytes.
(*d*) (i and ii) Diagrammatic cross-sections of a porocyte. (iii) Diagram of an entire porocyte of *Leucosolenia* with its central canal passing from the outside of the sponge (ext) to the choanocyte layer (c).
(*e*) Choanocyte of *Suberites massa*. The globular cell body with undulating areas of contact with adjacent choanocytes is typical of these cells in most sponges. The fringing collar (cl) of cytoplasmic tentacles and central long flagellum (f) are likewise constant features. In *Suberites* a special periflagellar membrane (pm) is developed between collar and flagellum. (Redrawn from Connes *et al.* 1971.)

with two exceptions, this has proved unrewarding. Feige (1969) has reported desmosome-like structures in *Ephydatia fluviatilis* and Ledger (1975) has observed septate junctions between sclerocytes in *Sycon*. There is, characteristically, a large area of overlap between adjacent exopinacocytes, where the cells are in close juxtaposition but the surfaces undulate. It is difficult to section this region in a plane perpendicular to the surface, sections are almost always oblique, which makes it difficult to be sure that apparent contacting surfaces are indeed specialized contact areas. The available evidence suggests that there is a gap of 10–30 nm between cells either at overlap or button junctions, and that there is little membrane specialization at points of contact. Although it is epithelial in function, probably phagocytic as well, the absence of any basal membrane distinguishes the sponge pinacoderm from the epithelia of other Metazoa.

A superficial coat of amorphous material invests the sponge external to the exopinacoderm, which can be greatly emphasized as a cuticle in some species.

Endopinacocytes (Fig. 2.1*c*)

Three characteristics distinguish endopinacocytes from exopinacocytes; their internal position lining all canals, the absence of a fibrillar superficial coating, and the fact that the nucleus is found in the same plane as the rest of the cell. Endopinacocytes are fusiform or subspherical, often with numerous thin extensions toward the mesohyl. There is always less overlap between adjacent cells than in the exopinacoderm and contact regions are thus very restricted. There is little membrane or cytoplasmic differentiation at contact points, and an intercellular space of 10–30 nm is usual. Both exo- and endopinacocytes make frequent contact with underlying myocytes, and the cellular network involving these cells provides the morphological basis for the contractile ability of sponges.

Table 2.1 *Cells of the lining layers (all dimensions in μm unless specified).*

	Overall shape	*Nucleus*	*Cytoplasmic structure*	*Vacuoles*	*Granular inclusions*	*Enzymatic activities*
Exopinacocytes	'T' shape or fusiform. In 'T' form 3.5–30.0 at the surface and 9–30.0 deep	Oval 1.3–3.0 × 1.7–4.0; nucleolus often absent; chromatin uniformly dispersed	Rough endoplasmic reticulum in cell body. Golgi marked. Lysosomes, primary and secondary. Mitochondria. Microfilaments and microtubules (20 nm diam.)	Phagosomes; triglyceride-containing vesicles	Glycogen	Acid phosphatase
Endopinacocytes	Fusiform to globular with cytoplasmic extensions – cell extremities very thin. Sometimes ciliated. Contact areas much reduced	Oval 3–4 long × 2–3 thick, nucleolus absent; chromatin dispersed	Rough endoplasmic reticulum and Golgi present. Mitochondria. Lysosomes. Microfilaments	Phagosomes; triglyceride-containing vesicles	Glycogen	Acid phosphatase

Basopinacocytes	Flattened 'T' shape forming a typical squamous epithelium; can produce filopodia	Oval with dispersed chromatin; nucleolus constant in fresh-water sponges only	Rough endoplasmic reticulum. Golgi marked, mitochondria abundant	Phagosomes in fresh-water forms. Large numbers of peripheral contractile vacuoles in fresh-water sponges	Not specified	Acid phosphatase located in granules not vacuoles
Porocytes	Cylindrical elongated, and produce filopodia	Oval, 2.6 × 1.9 chromatin aggregated into numerous patches. Nucleolus absent	Rough endoplasmic reticulum. Mitochondria. Lysosomes	Phagosomes contractile vacuoles in fresh-water forms	Present but character unspecified	
Choanocytes	Subspherical with flagellum and ring of cytoplasmic tentacles; occasional periflagellar membrane	Spherical 2–2.5 in diameter. Nucleolus usually absent	Blepharoplast; juxtanuclear Golgi; mitochondria dispersed	Phagosomes, pinocytotic vesicles. Vacuoles with paracrystalline inclusions		Proteolytic, lipolytic, amylytic

In terms of inclusions and cytoplasmic structure, endopinacocytes are similar to exopinacocytes.

Ciliated endopinacocytes occur frequently in the large exhalant canals of Dictyoceratida (e.g. *Hippospongia*) and in one sponge, *Oscarella*, ciliated exopinacocytes have been observed.

Basopinacocytes

By definition these cells form the basal epithelium of the sponge and are active in the secretion of the fibrillar collagen and polysaccharide complex, the basal lamina, which is the attachment structure for a sponge. In fresh-water sponges basopinacocytes are active in feeding, and extend filopodia into the surrounding water to engulf bacteria. They also are active in osmoregulation and contain large numbers of contractile vacuoles. In marine sponges the basal cells have a 'T' shape, are rich in mitochondria, but lack phagosomes and do not appear to retain the feeding role which their fresh-water counterparts show. Because of their activity in collagen synthesis, it is considered that basal pinacocytes represent a quite separate differentiation, not related to other pinacocytes but deriving either directly from archaeocytes, or representing collencytes specialized to secrete the particular fibrils found in the basal lamina. Some support for this idea derives from regeneration studies. Basal cells are slow to regenerate, implying that they must be supplied by differentiation of another cell and not simply replaced from adjacent exopinacoderm.

(b) Porocytes (Fig. 2.1*d*)

These specialized cells derive in the course of development from the exopinacoderm (Demospongiae) or the endopinacoderm of the oscular rim (Calcarea). They function as components of the inhalant system in sponges with very simple structure, where inhalant pores are surrounded by extensions of single porocytes and are thus designated as pore canals. The canal which traverses the porocyte in such simple situations places the external water in direct communication with the choanoderm. Porocytes are contractile and can open and close the pore in addition to regulating the diameter of the canal. There is a difference of opinion over whether the porocyte canal is intracellular or extracellular. In that the canal traverses the substance of the cell and is completely surrounded by continuous membrane, it is intracellular. However, as Borojevic (1966*a*) points out, the cell controls only the size of the canal, not the contents, and thus the canal cannot be intracellular in the same sense as a vacuole. In development the porocytes, which in Calcarea can at an early stage be differentiated from pinacocytes by their granular nature, elongate until they contact both exopinacoderm and choanoderm. The canal then opens at both ends and the porocyte membrane contacts choanocytes and exopinacocytes. Contact areas between these cells are unspecialized and the distance between opposing membranes is around 15 nm.

The cytoplasmic structure of porocytes is very similar to that of pinacocytes; cell contents include phagosomes, mitochondria and, in fresh-water sponges, contractile vacuoles. Although the cells are contractile no microfilaments have yet been observed in porocytes.

(c) Choanocytes (Figs. 1.1, 2.1*e*, 2.2)

These flagellated cells create the pressure which drives the water current and are always found lining choanocyte or flagellated chambers.

Choanocytes have irregular subspherical cell bodies closely applied to each other basally; the flagellum is apical, central and surrounded always by a circlet of about twenty cytoplasmic microvilli. In some species there is a periflagellar membrane which is a further apical extension of the choanocyte surface (Fig. 2.1*e*). The nucleus is central to basal and always lacks a nucleolus while the cell is functioning as a choanocyte. A prominent Golgi apparatus is present and is applied closely to the nucleus apicolaterally. This vesicle system is quite separate from the basal structures of the flagellum. Choanocyte cytoplasm is highly vacuolar as a result of active pinocytosis and phagocytosis.

Recent studies show the 'collar' tentacles to be connected by a mucous reticulum and freeze-fracture micrographs have revealed a core of microfilaments in each tentacle.

There is no basal membrane external to the choanocytes, the cells rest directly on the mesohyl and are held only by interdigitation of adjacent basal surfaces (Fig. 2.1). Choanocyte chambers vary in size and shape and in number of cells per chamber; however, they are always the easiest feature of sponge morphology to discern in sections, since they represent a definite aggregation of similar cells.

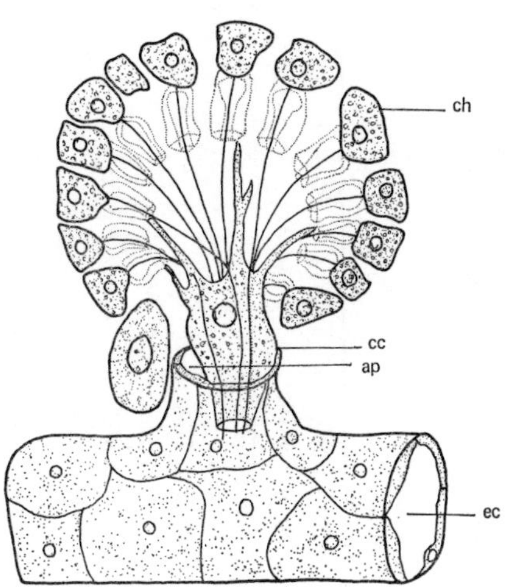

Fig. 2.2 A choanocyte chamber of *Suberites massa* showing the position and mode of operation of the central cell (cc). Choanocytes (ch) in this species have periflagellar membranes as well as collar tentacles. Pseudopodial extensions of the central cell are in contact with these membranes while the choanocyte flagella are trapped in canals in the cell body. The central cell also acts to reduce the diameter of the exhalant aperture (ap) leading to the exhalant canal (ec). (Redrawn from Connes *et al.* 1971.)

Table 2.2 *Contractile and skeleton secreting cells (all domensions in μm unless specified).*

	General shape and dimensions	*Nucleus*	*Cytoplasmic structure*	*Vacuoles*	*Granular inclusions*	*Enzymatic activity*
Collencytes	Stellate fusiform with irregular filopodia	Oval, central; nucleolus often absent	Normal granular reticulum; mitochondria	Collagen inside vacuoles already showing periodic structure	Glycogen	
Lophocytes	Ovoid cell body always with numerous filopodia – around 20 longest dimension. Always with dense collagen 'tail'	Oval, nucleolus present	Complex with many inclusions, microtubules, moderate endoplasmic reticulum. Mitochondria small and numerous	Osmiophilic vesicles, lipid filled vacuoles. Phagosomes intracellular collagen sometimes detectable		
Spongocytes	Cells subspherical with few extensions. Polarized with relation to the object around which they are secreting	Spherical, nucleolus present	Very marked development of rough endoplasmic reticulum; cytoplasm vesicular	Vacuoles involved in collagen aggregation, osmiophilic		

Sclerocytes	Cell spherical, lacking filopodia	Spherical, nucleolus present	Well-developed endoplasmic reticulum abundant. Mitochondria abundant. Cytoplasmic microtubules and vesicles	Large clear vacuoles at each end of spicule rudiment		
Myocytes	Fusiform around 50 × 2.3 with irregular cytoplasmic extensions toward the extremities	Oval, central 1–2 wide × 2–4 long, nucleolus present or absent	Reticulum weak. Mitochondria present. Lysosomes. Microtubules and (15–25 nm) microfilaments (5–7 nm) abundant	Phagosomes; triglyceride-containing vacuoles	Glycogen	Acid phosphatase. Cholinesterase

2.2.2 Cells which secrete the skeleton (Table 2.2)

(a) Collencytes (Pl. 1*b*)

The fundamental component of the sponge skeleton is the fibrillar collagen which is the framework for the entire sponge matrix. The cells whose primary function is to secrete this material are termed collencytes. These cells are, morphologically, very close to pinacocytes and in cell suspensions when all cells become spherical the two types cannot be distinguished. In the parenchymella larva of Demospongiae, most of the internal cells are capable of secreting collagen, an activity which attentuates or ceases when the cell takes up a superficial position. The first definitive collencytes are those cells which remain internal after a pinacoderm has been organized during metamorphosis. These cells are stellate or fusiform, with an oval central nucleus which frequently lacks a nucleolus. There are few cytoplasmic inclusions, but vacuoles which apparently contain recognizable collagen are often found (Pl. 1*b*). The secretory activity of these cells is evident from the dense depositions of oriented collagen fibrils in the mesohyl adjacent to the cell, and sometimes attached to the cell membrane.

(b) Lophocytes (Pl. 2*a*)

These are large, very mobile cells which occur in greatest number just above the base of the sponge. They are recognizable by the production of great quantities of fibrillar collagen which is secreted behind the cell as it moves. At any given moment, a band of collagen remains attached to the cell membrane and moves as a tail following the cell (Pl. 2*a*). In some species (e.g. *Chondrosia*) lophocytes are extremely irregular cells with filopodia around most of the surface. In terms of function, lophocytes are very close to collencytes; however, in addition to the collagen tail, they differ from the latter in having extremely complex cytoplasmic structure and usually in possessing a nucleolus. It is probable that the emphasis placed by sponge cytologists upon the presence or absence of a nucleolus when attempting to establish relationships between cell types is quite misleading, the lophocyte–collencyte relationship being a case in point.

Lophocytes have not been reported to occur in all groups of sponges, but certainly they occur in Calcarea, fresh-water sponges and Hadromerida. On the basis of time-lapse cinephotographic observation, Efremova (1967) suggested that lophocytes were merely one particular form assumed by mobile amoeboid cells when they approached the sponge surface. With their marked polarity, fast movement, and fibrillar tails, lophocytes are a striking feature of the superficial region of *Ephydatia fluviatilis* and the evidence from ultrastructural studies in this and other species, favours the view that they are a stable feature, occurring in the same form wherever heavy collagen deposition is necessary.

(c) Spongocytes (Pl. 1*a*)

These cells elaborate the perispicular or fibre-forming collagen of the type traditionally referred to as spongin. This constitutes the major binding element of the skeleton in several orders of Demospongiae. Spongocytes are always observed to operate in groups, clumping around spicules or around a growing fibre, and can be characterized by this co-operative behaviour, which is no doubt necessitated by the large quantities of material the cells must produce to stabilize the skeletal network. The resulting cell aggregations are visible even in light microscopy.

Morphologically, spongoblasts are characterized by the presence of an extensive rough endoplasmic reticulum, by their polarization, always being wrapped around or applied to a spicule or a fibre (Fig. 3.2, p. 89), and by a vesicular cytoplasm which contains dark homogeneous vacuoles of collagen precursor material.

Studies using autoradiography to monitor collagen synthesis by means of proline incorporation have provided a very clear demonstration of the collagen synthesizing activities of these cells in *Ephydatia mulleri* and *Haliclona elegans* (Garrone and Pottu 1973).

(d) Sclerocytes (Fig. 2.3; Pl. 4*b*)

A sclerocyte is a mobile mesohyl cell which completes or contributes to the formation of a spicule. There have been few studies of the sequence of spicule development, but deposition of silicon is known to take place around an axial proteinaceous filament within a vacuole, or extending between two vacuoles at opposite ends of the cell. The membrane enclosing the axial filament, the silicalemma, has the function of transporting or pumping silicon. The sclerocyte has well-developed rough reticulum, prominent nucleolus, abundant mitochondria, cytoplasmic microfilaments and many small vacuoles.

In *Haliclona*, the axial filament protein has been studied biochemically and with electron microscopy, and has a crystalline substructure and an overall hexagonal outline.

After spicule secretion is complete, sclerocytes disintegrate.

2.2.3 Cells which cause contraction (Table 2.2)

(a) Myocytes (Pl. 3)

These contractile elements of the sponge mesohyl are characterized by their overall fusiform shape, which is often distorted by long projections, by their localization, grouped in concentric fashion around oscules and major canals, and by the great number of microtubules and microfilaments contained in their cytoplasm and disposed parallel to the long axis of the cell. The dimensions of these tubules and filaments accord with those described in mobile and contractile cells of other organisms. The fact that large, 1500–2500 nm, and small filaments, 500–700 nm, occur, and that the former predominate, allies myocytes with smooth muscle cells of invertebrate type. Myocytes are the only sponge cells which give a positive reaction to tests for cholinesterase activity.

It is cells of myocyte type, with an unquestioned contractile function, that,

Plate 1

(a) Electron micrograph of a group of three spongocytes (sp) wrapped around two spicules (s) and secreting the material which will form the perispicular spongin (ps). The dense inclusions (i) in the cells are the collagen precursor material which is released in packets to contribute to spongin formation. Residual concentrations of precursor material are visible in the developing spongin (pm). (From Garrone and Pottu 1973.)

(b) Electron micrograph of a portion of a collencyte showing the general fusiform aspect of the cell and in this case, intracellular vacuoles (iv) containing collagen in which periodicity is already apparent. Dense inclusions of unorganized precursor material (pm) are found throughout the cell. The dense collagen of the mesohyl matrix is well shown in this photograph.

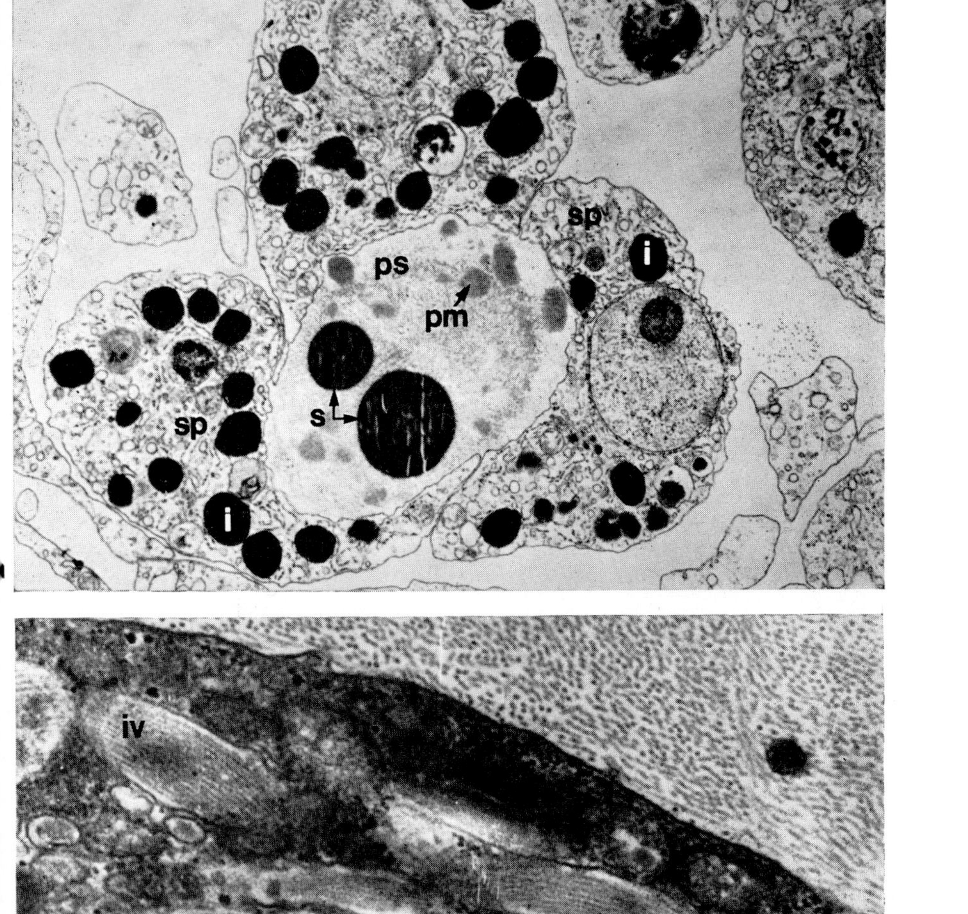

iv
iv
pm

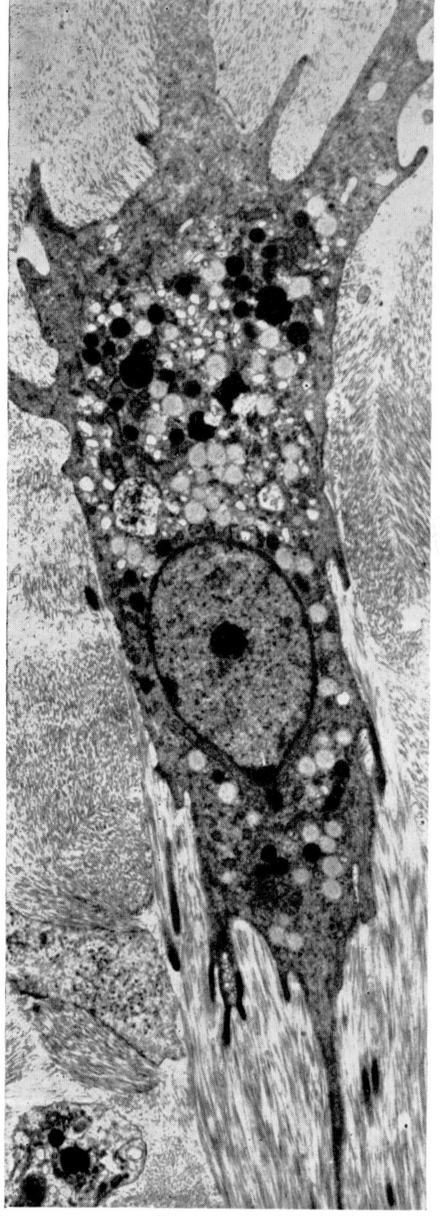
a

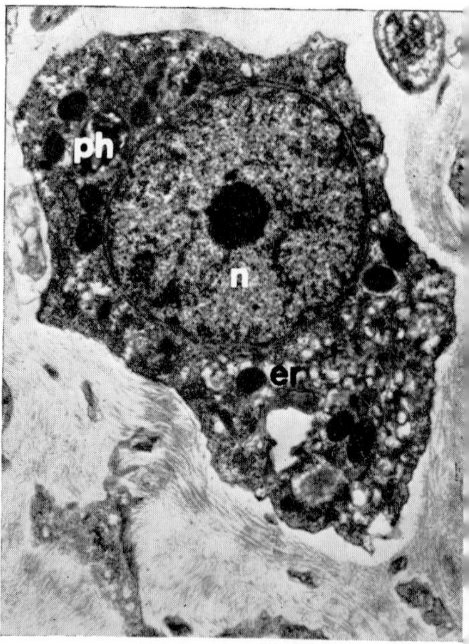
ph
n
er

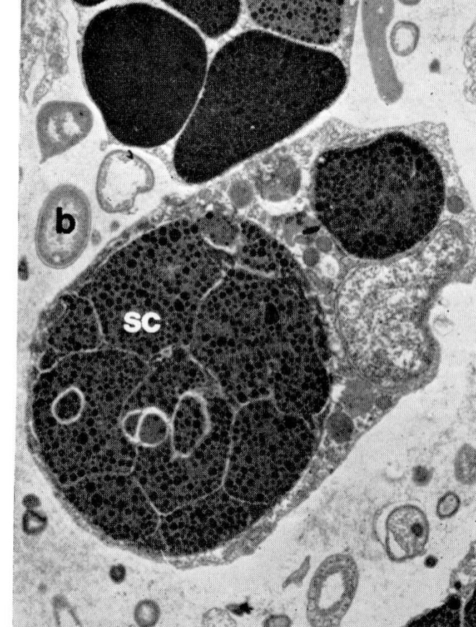
b
sc

Plate 2

(a) Longitudinal section through a lophocyte showing the complex vesicular cytoplasm, nucleolate nucleus, multiple surface projections and the prominent 'tail' of collagen which trails behind this mobile cell as it moves through the matrix. (From Garrone 1971.)

(b) A typical archaeocyte with large nucleus (n), prominent nucleolus, tubular endoplasmic reticulum (er), mitochondria and phagosomes (ph) in several stages of digestion.

(c) An archaeocyte engaged in phagocytosis of a spherulous cell (sc). The symbiotic matrix bacteria (b) which occur in many sponges are also shown.

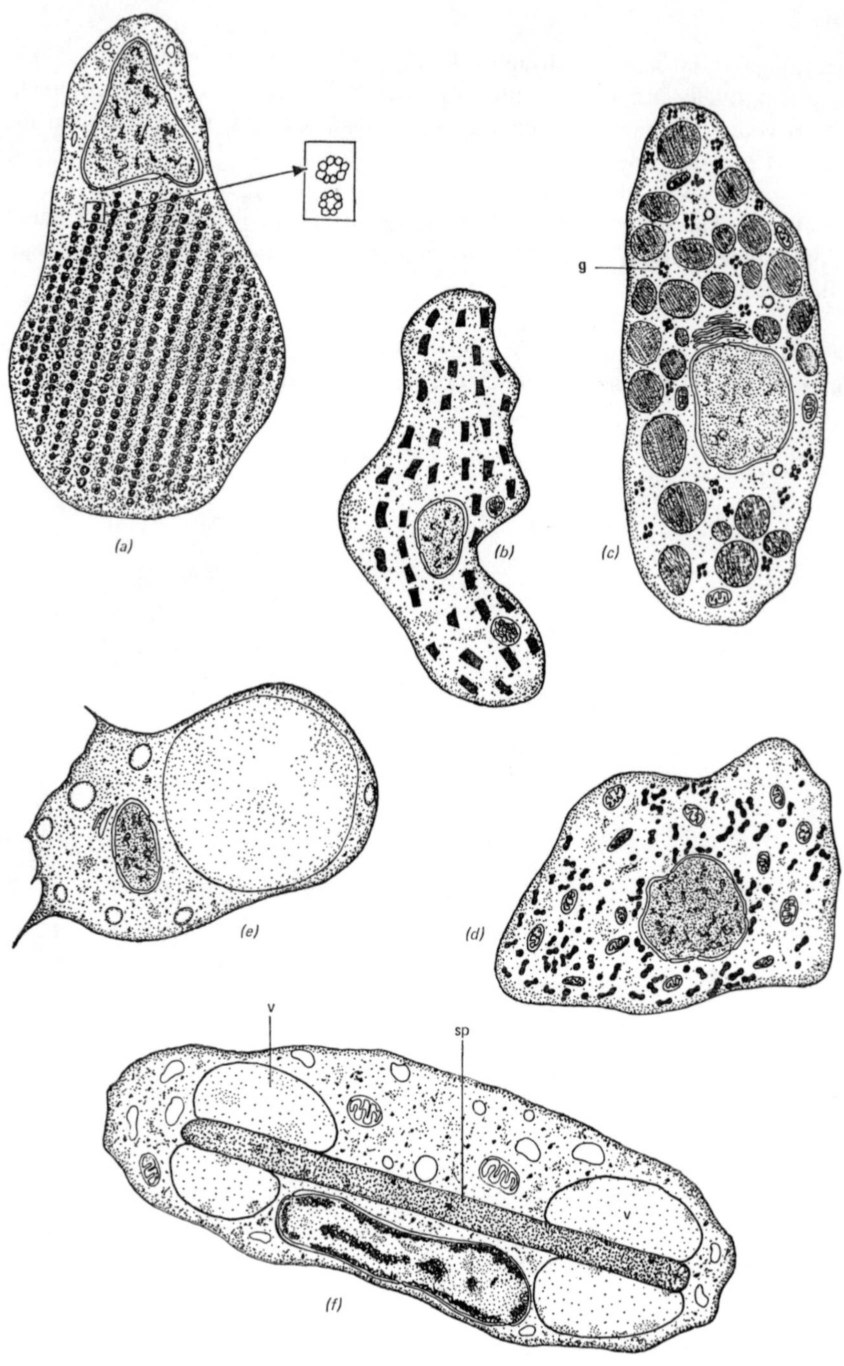

g
(a)
(b)
(c)
(e)
(d)
v
sp
v
(f)

Fig. 2.3 Sclerocyte and cells with inclusions.
(*a*) A globoferous cell of *Ophlitaspongia. Inset*: structure of the paracrystalline bodies.
(*b*) A rhabdiferous cell of *Microciona* with dispersed rod-like inclusions.
(*c*) A gray cell of *Ophlitaspongia.* Glycogen granules (g) and dense basophilic vesicles characterize cells of this type.
(*d*) A microgranular cell of *Ophlitaspongia.* The approximate spherical shape and tiny homogeneous inclusions are constant features of this cell type.
(*e*) A cystencyte of *Ephydatia mulleri.*
(*f*) A sclerocyte of *Mycale* with a rudimentary spicule (sp) extending between two vacuoles (v).
(All redrawn from electron micrographs.)

since the early 1950s, have been the focus of a continuing debate on whether sponges possess a system, either structurally or functionally, analogous to the nervous system of other Metazoa. Early cytological work using silver impregnations of sponge tissue, was certainly open to question when used to provide evidence for the existence of cells of nervous type. The techniques were not specific, and far too little was then known of histology and cell function in sponges. However, since the initial histological work which prompted the suggestion that nervous, neuromuscular and sensory elements could be demonstrated in a number of Demospongiae and Calcarea, Pavans de Ceccatty and his co-workers have gone on to provide excellent evidence for the existence of an elementary integrative system in sponges. The largely semantic debate pursued by his critics, notably Jones (1962), is now irrelevant. Many of the points quoted by Jones as indications that cells of muscular and integrative type do not occur in sponges can now be used to support the argument for the presence of these cells. For example, acetylcholine and cholinesterase do occur in sponges (Lentz 1966; Thiney 1972), and myocytes do possess microfilaments (Thiney 1972). At this point in time Jones's review should be read for its excellent coverage of early observations on the histology and physiology of sponges.

It is clear that much misunderstanding can be avoided if one does not refer to a sponge as possessing a 'nervous system' but, rather, emphasizes that the necessary attributes of such a system are largely met. Specifically, cells of myocyte type are organized in a network contacting at the points of filopodia with other myocytes and with pinacocytes (Pl. 3; Fig. 2.4). These cells have pacemaker activity as demonstrated both by photographic means and by noting their response to certain chemical stimuli. Contractions are dispersed from cell to cell very quickly, that is to say the system can conduct stimuli and, further co-ordinated rhythmic activity can occur (Reiswig 1971*a*).

However, here the analogy must stop. All attempts to detect action potentials in sponges have failed, and nothing approaching synaptic connections as seen in higher organisms is known. The failure to detect action potentials may reflect

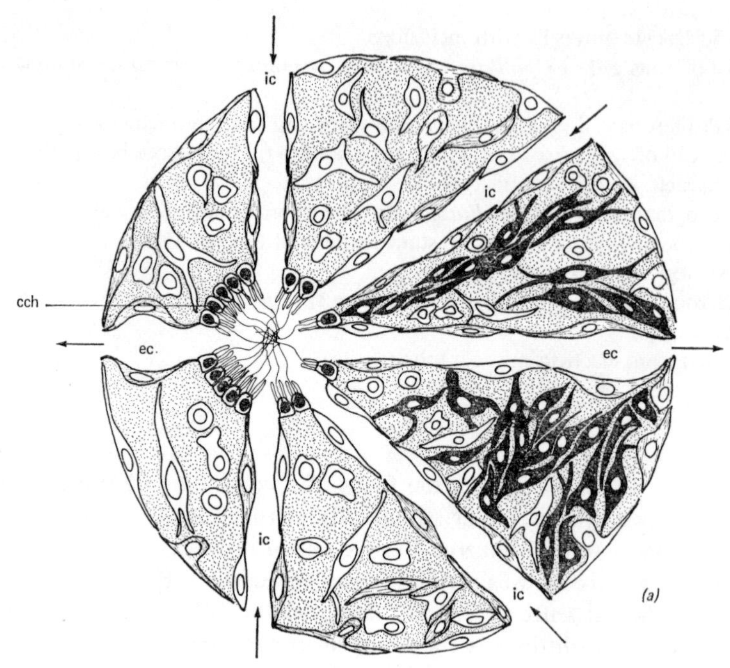

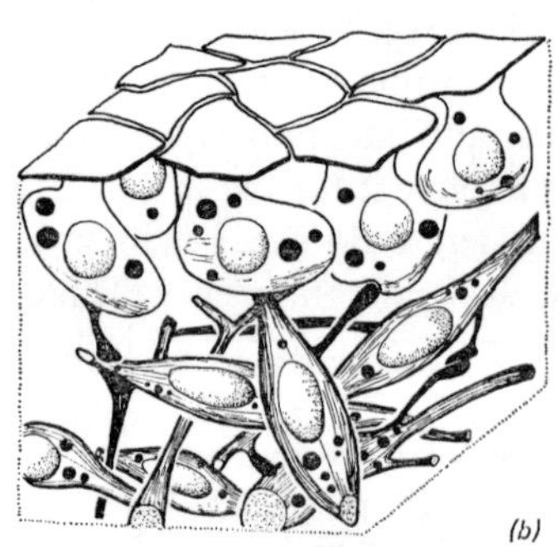

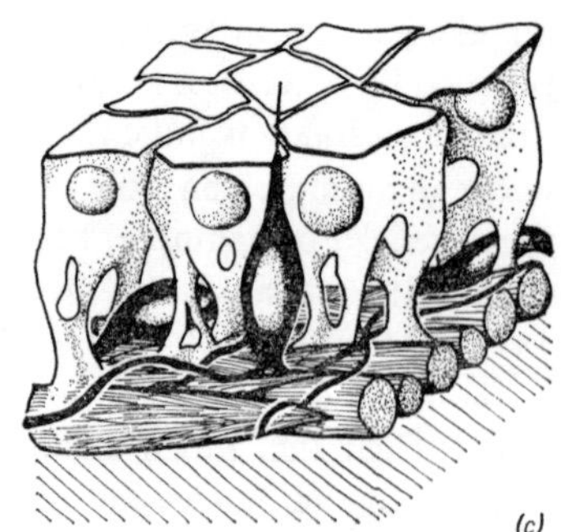

the inadequacy of our present methods for monitoring something very weak or very slowly generated; at present we just do not know.

Because of these very real differences between the integrative elements in sponges and those seen in Coelenterata, and also because of the mobility and instability of all sponge cells, it is better to refer to the sponge as having an integrative or protonervous system. The former term is preferable because it does not emphasize a distinction between what is seen operating in sponges and the smooth muscle and epithelial conductive systems of other invertebrates.

Fig. 2.4 Diagrammatic representation of the integrative network in sponges and cnidarians.
(*a*) Diagram of the supposed co-ordinative contacts in sponges. The direction of the water current is indicated by arrows. The three sectors represented from left to right portray the varying architecture of sponges, particularly with respect to mesohyl volume. Thin-walled species are to the left; thick-walled species to the right. cch, choanocyte chamber; ic, incurrent canal; ec, excurrent canal.
Several co-ordination or transfer pathways are possible:
(i) Fluid extracellular pathways for substances spreading in the mesohyl, the dimensions and macromolecular density of which increase in thick-walled species.
(ii) Mobile cellular pathways involving transitory contacts of wandering cells; these become more localized in thick-walled sponges.
(iii) Fixed pathways, where, in sponge terms, relatively stable relationships are established between myocytes and epithelia, or myocytes and other mesohyl or choanocyte elements. Such contacts are reinforced by areas of junctional specialization. Epithelial networks are common, deeper systems appear only in thick-walled species.
(*b*) Diagram of the connection between sponge pinacoderm and contractile myocytes of the underlying mesohyl. There is no basement membrane.
(*c*) Diagram of the myoepithelium and nervous system of a cnidarian. The muscle sheet is composed not of individual cells, but of contractile processes arising from the base of epithelial cells. Special sensory cells run in narrow spaces between the epithelial cells. A basement membrane supports the whole structure.
(*a*, after Pavans de Ceccatty 1974*a*; *b*, *c*, after Pavans de Ceccatty 1974*b*.)

Recent work on scyphozoan conductive epithelia has extended our concept of simple conducting systems. A myoepithelial network such as that in sponges, albeit unstable and without basement membranes separating the tissues, certainly becomes comparable to conducting networks of the Cnidaria.

2.2.4 Totipotent amoeboid cells

(a) Archaeocytes (Pl. 2*b*, *c*)
Archaeocytes are the indispensable cells in a sponge, it is generally considered that they are capable of undergoing any kind of differentiation and thus can give rise to any other cell type. Archaeocytes are large cells, very mobile, with large vesicular nucleus and prominent nucleolus. Clearly they have a major role in digestion, since their cytoplasm is always charged with phagosomes. Rough endoplasmic reticulum is present and shows a tubular type of organization; mitochondria are abundant and all stages of the lysosomal cycle can be identified.

In the mesohyl archaeocytes are always dispersed, except in cases of a special activity such as gemmulation, budding or repair after injury. Archaeocytes have been shown to possess acid phosphatase, protease, amylase and lipase activity.

These cells are responsible for any phagocytosis which takes place through the pinacoderm of canals, for phagocytosis of matrix bacteria or expended cells (Pl. 2*c*) and also for accepting phagocytosed material from the choanocytes.

The term 'amoebocyte', often used with reference to sponge cells, has no precise meaning. It is used merely to denote any mesohyl cell where no special activity can be determined; an amoebocyte with multiple phagosomes would be termed an archaeocyte.

It is necessary to define an archaeocyte in two ways. First, from the functional viewpoint, the archaeocyte is the major phagocyte, the macrophage, of the sponge and, as such, carries out the bulk of digestive and excretory activity. Secondly, from the point of view of flexibility in development and differentiation, archaeocytes are the only cells which are absolutely necessary to the formation of a functional sponge as is demonstrated by the development of sponges from gemmules which have a uniform archaeocyte population. Because of this demonstrated totipotency in the course of gemmulation, it is assumed that archaeocytes, at any time, in larval or adult sponge, retain this capacity to differentiate in any direction. What is very hard to assess is the extent to which differentiations are reversible, at least morphologically, or the extent to which cell types derived from archaeocytes can differentiate further. Can a collencyte become a lophocyte or a myocyte? We will examine this question in relation to sponge morphogenesis.

2.2.5 Cells with inclusions

Under this general heading are grouped all mobile mesohyl cells which can be characterized by the presence of granules or vesicles of a particular morphological or chemical type. These cells, which can be described very precisely, are derived from archaeocytes directly and represent terminal differentiations.

So far no indisputable function has been ascribed to any of the following cell types except rhabdiferous cells. Also, it is now clear that the same term has sometimes been used to describe cells in different orders of Demospongiae in advance of any demonstration that the structure and function are equivalent.

(a) Spherulous cells (Figs. 2.5*d*, *e*; Pls. 2*c*, 4*a*)

Spherulous cells, as the name suggests, are filled with large round vacuoles which ultimately come to occupy almost the entire cell, thus compressing the cytoplasm into thin strands which run between the nucleus and the cell boundary. The cytoplasm contains little in the way of inclusions, although lysosomes persist until a late stage of vacuole expansion. Spherulous cells have been described from many Demospongiae and are always most abundant superficially and bordering exhalant canals. In some sponges the concentration of these cells below the surface marks a distinct tissue. While the spherulous cells described from different groups have the same vesicular structure and the same localization, the nature of the vesicle contents is very different from group to group. For example, those described in *Tethya* and *Suberites* are very distinct from spherulous cells in *Verongia* and *Haliclona*. It is certainly not clear that all cells termed spherulous are equivalent in function. The problem of characterizing the inclusions and of understanding the developmental sequence through which the cells pass remains unresolved. The development of spherulous cells is best understood in *Verongia*, and here they are frequently extruded from the sponge into the exhalant canals and it is possible that the chromolipoidal material they contain is an excretory component. However, the cells are present in large numbers and certainly do not only serve as reservoirs of material for excretion. In the living sponge the cells are bright yellow-brown to brown and highly refractile.

It is possible to trace the development of these cells from archaeocytes, and the observation that they are frequently extruded suggests that the differentiation is not reversible; further, phagocytosis of entire spherulous cells by archaeocytes is common (Pl. 2*c*).

(b) Microgranular cells (Figs. 2.3, 2.5*a*, *b*)

These are cells which are usually approximately spherical, dispersed and found in low numbers in the mesohyl. They are characterized by a cytoplasm filled with small dense granules. Mitochondria are present in considerable numbers and the granular endoplasmic reticulum is moderately developed. The nucleus is smaller in relation to cell volume than in archaeocytes. In the case of micro-granular cells, it is clear that two distinct types of cell, as judged by granule size and structure have been grouped under this name.

In Poecilosclerida and Hadromerida the term has been applied to cells where the numerous granules are small rods (0.3–0.5 by 0.1–0.2 μm wide) in which the contents are homogeneous (Fig. 2.3*d*).

In Dictyoceratida, microgranular cells contain fewer granules which are large, dense, osmiophilic structures, having a secondary highly organized sub-structure (Fig. 2.5*a*, *b*).

The function of both types of cell remains unknown.

(c) Gray cells (Figs. 2.3*c*, 2.5*c*)

These are best characterized by reference to poecilosclerid sponges belonging to the family Clathriidae from which they were first described, and where they constitute around 4 % of the cell population in suspensions.

In cell suspension, gray cells are very mobile and continue vigorous move-ment even after an aggregate has formed. Gray cells can be easily detected by this mobility in conjunction with a cytoplasm charged with large, oval to spheri-cal, dense, basophilic granules and tiny granules of glycogen. The cells are ovoid, the nucleus small, the reticulum is not well-developed and mitochondria are rare.

The contents of the large granules appear slightly fibrillar at high magnifica-tion and two components of unequal density can be discerned.

In Dictyoceratida and in *Chondrosia*, cells of somewhat similar type occur

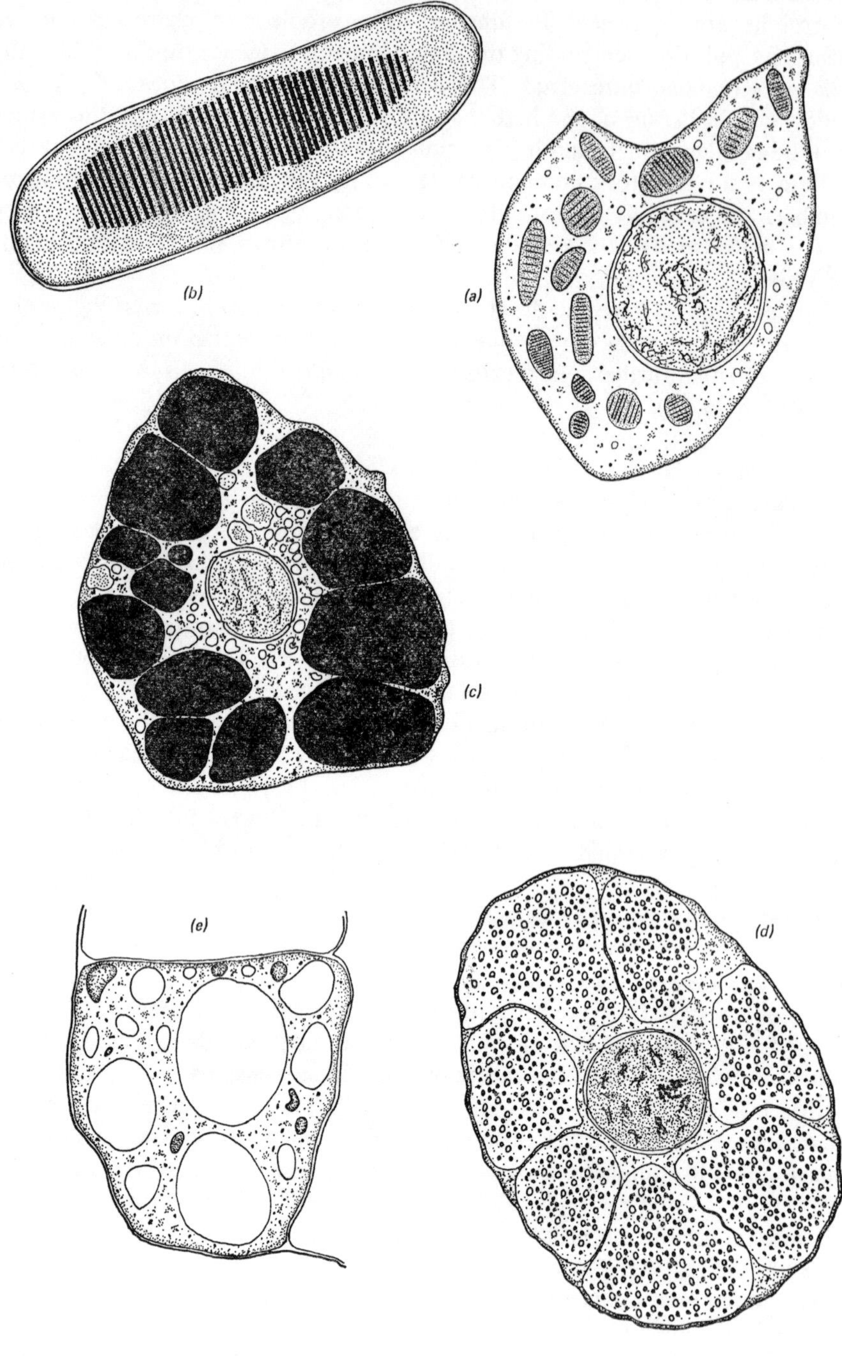
(b)
(a)
(c)
(e)
(d)

Fig. 2.5 Cells with inclusions.
(*a*) Microgranular cell of *Pseudoceratina.* The ovoid inclusions with paracrystalline central structure are characteristic.
(*b*) Detail of one granule of (*a*).
(*c*) Gray cell of *Pseudoceratina.* The cytoplasm is more compressed by the basophilic vesicles than in the gray cell shown in Fig. 2.3*c*.
(*d*) Spherulous cell of *Ianthella.* Greatly reduced cytoplasm and large vacuoles which contain granules of two distinct types are the notable features of these secretory cells.
(*e*) An early stage in the development of one vacuole of a spherulous cell. The inclusions are larger and large spaces are still found within each vacuole.
(All redrawn from electron micrographs.)

and have been termed gray cells, but again no histochemical correspondence between these and poecilosclerid gray cells has been established. In *Pseudoceratina* and *Verongia* the basophilic inclusions are few in number and fill the entire cell leaving only a central nucleus and thin strands of cytoplasm in a manner similar to that of the spherulous cells.

(d) Globoferous cells (Fig. 2.3*a*)
These cells are easily recognizable by their small conical nucleus, displaced to one end of the cell by the expansion of a single large globule. Cytoplasm is reduced to the perinuclear area and the cell periphery. The ultrastructure of the globular area is highly ordered, consisting of parallel rows of cylindrical rods which become increasingly disordered until they grade at the periphery into the cytoplasm. Each cylinder is hollow, lined by around twelve longitudinal elements. No histochemical tests done so far have characterized these paracrystalline components, and cells of this specific type are known to occur only in poecilosclerid sponges.

(e) Rhabdiferous cells (Fig. 2.3*b*)
These are large cells, oval or fusiform, and containing numerous rod-like inclusions dispersed through the entire cytoplasm and aligned parallel to the long axis of the cell. The inclusions in this case are acid mucpolysaccharide, and, since rhabdiferous cells have been often observed to discharge their contents into the mesohyl, it is likely that they elaborate and store at least some of the polysaccharide ground substance of the sponge matrix. Rhabdiferous cells discharge by disruption of the entire cell and are clearly a terminally differentiated cell state. At discharge, the inclusions become diffuse and reticulated but show no change in chemical properties.

(f) Cystencytes (Fig. 2.3)
Known only from fresh-water sponges, these cells are distinguished by having a single large vesicle containing amorphous material and occupying most of

the cell. The cytoplasm around the vesicle is reduced to a thin film but becomes more extensive around the nucleus. Golgi apparatus, rough and smooth endoplasmic reticulum and contractile vacuoles are present. The secretion in the cystencyte vacuole is polysaccharide and, conceivably, these are functional equivalents of the rhabdiferous cells of other Demospongiae.

2.3 Cell behaviour in aggregation

The remarkable ability of sponges to aggregate after mechanical dissociation was discovered early this century by Wilson, and the phenomenon was the subject of much study up to 1930. The far-reaching implications of Wilson's observations for the field of developmental biology did not become apparent for several decades. Meanwhile, interest in the regenerative behaviour of sponges lapsed, and their aggregation ability was regarded as merely another peculiarity of these organisms. However, recent work on sponge aggregation, predominantly by Humphreys and Curtis, interpreted in terms of modern knowledge of cell surface structure, has proved to be of great significance in relation to the basic question of how cells adhere and segregate.

Detailed review of the numerous experiments on sponge aggregation is not necessary here, a brief summary of the facts will suffice to emphasize both points of significance, and the problems which remain. Much of the experimental information relates to *Microciona prolifera*, the common red beard sponge of the Atlantic coast of North America, but almost any sponge dissociated and sustained under proper conditions will form aggregates, and many will eventually reconstitute an aquiferous system.

If cells of *M. prolifera* are pressed through fine cloth into freshly made up sea water, the resulting suspension will contain individual cells and clusters of cells. When allowed to stand, the cells migrate actively, and aggregate instantaneously. Finally, after two to three weeks, they form a functional sponge with canals, spicules, and collagen fibrils and fibres. Similar behaviour has been observed with many species, the main variable being the rapidity with which aggregation proceeds.

A further observation, also made by Wilson, was that if cell suspensions of two different-coloured sponges were mixed, in his experiment *Haliclona oculata* (purple) and *M. prolifera* (red), aggregation proceeded to the formation of separate sponges. Small *Haliclona* and *Microciona* reconstituted out of the cell mixture. Aggregation thus appeared to be species specific.

This concept of specificity of cellular adhesion, which arises primarily out of work on sponge aggregation, has had a great influence on thinking and experimentation in animal morphogenesis. Whether or not a specific molecular adhesive mechanism operates is, however, still in doubt, and thus two viewpoints must be considered.

The case for the occurrence of specific adhesion rests mainly on the work of Humphreys (1963, 1970). Building on earlier observations, Humphreys noted that mechanically dissociated cells aggregated rapidly at 24°C and only slowly at 5°C. He wondered what factors were involved in the reconstruction of a sponge and thought, in the light of his observations, that it was possible that living sponge cells synthesize a species-specific product which acts at the cell surface and which is essential for the formation of an initial adhesion after cells collide.

To test this idea it was necessary to design standard experimental procedures. First, to enable consistent dissociation of the sponge into single cells the tissue was soaked for a few hours in calcium- and magnesium-free sea water. This chemical treatment was gentle enough to break bonds between the macromolecules which are involved in cell adhesion, but not harsh enough to destroy the molecules. Cells dissociated in this way aggregate exactly as do mechanically dissociated cells. A second major problem in studying cell aggregation on glass is to distinguish between migratory and adhesive properties of cells. These must be brought together in consistent manner so that the size, number and rate of formation of aggregates can be related directly to adhesive properties, not to cell movement. Use of a gyratory shaker at a constant rate of rotation for the aggregation experiments overcame this problem.

Chemically dissociated cells of *M. prolifera* and *H. oculata* behave as follows. When kept in sea water free of divalent cations, they remain dissociated. When returned to normal sea water at 24°C, they reconstitute a species-specific aggregate, thus behaving exactly as do mechanically dissociated cells. However, if they are returned to normal sea water at 5°C, they do not aggregate. The conclusion therefore is that aggregation of sponge cells requires both divalent cations, and the operation of a temperature-dependent process which does not appear to operate at 5°C.

The inability of cells to aggregate at the lower temperature can be reversed by the addition of supernatant from the dissociation medium to the normal sea water in which the cells are suspended. This activity in the supernatant indicated the presence of a material which had been removed from the cells chemically, since cells dissociated mechanically retain their ability to aggregate even at low temperatures. In the absence of any knowledge of the nature of this material it was termed by Humphreys 'aggregation factor' or 'factor'.

Factor, in the *M. prolifera/H. oculata* mixed experiments, acts in a species-specific manner. When *Haliclona* factor is added to an interspecific mixture only *Haliclona* aggregates form, addition of *Microciona* factor will bring about adhesion of *Microciona* cells. It should be noted here, since criticism has been made of the interpretation placed on these observations, that *M. prolifera* factor, while species-specific in combination with *Haliclona oculata*, behaves non-specifically with *Halichondria panicea*, a sponge with which *M. prolifera* will initially form finely mixed aggregates. However, the existence of reciprocal specificity in the *Microciona/Haliclona* system has allowed this to serve as a model for further hypothesis and experiments.

It is necessary to separate observation and interpretation before completing this synopsis. There is clear evidence that sea water in which *M. prolifera* and *H. oculata* are chemically dissociated contains a product that can promote species-specific aggregation when added to cell suspensions held at 5°C. This is interpreted to mean that the product is synthesized at 24°C but not at 5°C. It was, however, premature to conclude from these experiments that this 'factor', which is released by the action of calcium- and magnesium-free sea water, resides normally at the cell surface and plays an active and initial role in establishing normal relations between sponge cells.

Recently the aggregation factors from two sponges have been characterized (Henkart, Humphreys and Humphreys 1973; Muller and Zahn 1973). Factor derived from *Microsciona parthena* is a large acidic proteoglycan complex which sediments at 70 S and is composed of 47 % amino acid and 49 % sugar. Electron microscopy of purified factor shows that the complex is fibrous, with a central globular core, radiating arms and an overall diameter of 160 nm. When calcium ions are removed from factor solution the complex dissociates into glycoprotein subunit molecules each of 2×10^5 daltons, plus a core. In *Geodia cydonium* the aggregation process occurs in three clearly distinct steps, and the suggestion is that the primary and secondary phases are under the control of separate aggregation factors. The molecule which promotes secondary aggregation has been purified by Muller and Zahn and it is a large annular proteoglycan particle with a contour length of 350 nm. From the core, about twenty-five filamentous arms radiate. The aggregation-promoting activity resides in the arms, which are entirely protein. The core appears to function as a spacer or carrier molecule. The carbohydrate component of *Geodia* factor is very much lower than that of *Microciona* factor. With the structure and composition of factor now known, it should be possible to locate large molecules of this type in membranes or cell coat material of sponges and thus better understand its site and mode of action.

Curtis has for a long time held that adhesion of sponge cells is not specific; that there is no evidence which requires the intervention of specific molecules at, or between, cell surfaces when cells collide and then adhere. Recent observations on the activity of aggregation factors in various strains of *Ephydatia fluvialitis* (Curtis and Van de Vyver 1971) have allowed the formulation of a hypothesis which in some measure reconciles what has appeared to be an irreconcilable divergence of opinion. Factors are released from dissociated cells of two strains, alpha and delta, of *E. fluviatilis*. Alpha factor increases the adhesiveness of alpha cells and decreases the adhesiveness of delta cells, and vice-versa for delta factor. Further, the extent to which adhesive capacity is reduced in mixed alpha–delta suspensions is related directly to the respective concentrations of the factors in the suspensions.

If we consider what happens when two sponges of different strain, or species, contact, we see that the opposed pinacoderms initially form an adhesion, but the cells do not interpenetrate. After about 24 h in *E. fluviatilis* the adhesion collapses and the sponges of each strain separate. When the contacting sponges are of the same strain, or in marine sponges the same species, the adhesion forms, the contacting pinacoderm cells migrate to the surface of the fusing sponges and the mesohyl components interpenetrate.

If one assumes that cells show complete specificity of adhesion, as in the *Microciona prolifera*/*Haliclona oculata* experiments, then sponges of unlike strains should not form even temporary adhesions.

The main features of this coalescence or non-coalescence of different strains or species can be explained in terms of control by the alpha and delta factors. In sponge alpha, the concentration of alpha factor will be highest at the body centre and will decrease toward the periphery. Therefore, adhesiveness of the surface cells of the sponge body will be low, but not negligible. When two sponges of like type contact, an adhesion forms. Since the factors are diffusible and may act without irreversible binding to the cells, the concentration of the factor will rise in the region of contact between sponge bodies. As the homologous factors increase in concentration, cells become more adhesive and a single permanent sponge body will be established.

When two unlike sponges contact, the initial adhesion forms because the concentration of the two factors is low at the surface. Shape changes along the region of contact will allow the concentration of both factors to rise by preventing surface diffusion. Cells exposed to the heterologous factor will become non-adhesive and this loss of adhesive capacity will be greatest at the two contacting surfaces. Hence the sponges cease to adhere.

These observations allow several differing interpretations to be reconciled.

Humphreys identified a factor which specifically promoted the adhesion of cells of the species type from which the factor was derived. He did not test to discover whether the factor diminished the adhesiveness of cells of other species, and he assumed that the factors act as cements which attach to binding sites on two apposed cell surfaces. No evidence exists that the factors are absorbed by cells. Many other experiments, using different species, have suggested that specific cell adhesion does not occur and, when differing experimental procedures have been allowed for, the objections raised by Curtis (1970) to the specificity case are the strongest, and need answering.

A reinterpretation of these conflicting results in terms of the above hypothesis which stems from the *Ephydatia* experiments has been advanced by Curtis and Van de Vyver (1971). Assume that the factors discovered in *E. fluviatilis* strains have species-specific counterparts in marine sponges, and thus that the factors isolated from *Microciona* and *Haliclona* are identical in general behaviour with those in *E. fluviatilis*. They could act not as agents of specific adhesion, but as substances controlling the appearance or loss of an adhesion, the direct mechanism of which would be unspecific. In the experimental design used by Humphreys, factors of the *Ephydatia* type would give exactly the results he obtained. That is to say aggregation of cells of a heterologous type would be prevented, and adhesion of cells of homologous type promoted, in the presence of a single factor. When two species types are aggregated in the presence of both factors, it would be expected that the initial adhesions would be random. Later, as small groups of cells of one species type began to produce appreciable amounts of their factor, they would make the local environment less favourable for interspecific adhesions and more favourable for intraspecific adhesions. Thus sorting out of species would occur.

If the above interpretation is accepted, then apparently contradictory results on sponge cell adhesion can be reconciled. Diffusible factors are produced, but their function is specifically to control adhesiveness of cells whose mechanism of adhesion itself is unspecific.

Research on sponge aggregation factors and the aggregation process itself is in a very active phase at present and any final statement on the subject would be premature. Two very significant recent approaches have been made to different aspects of the problem. John, Campo, Mackenzie and Kemp (1971) have attempted Ficoll gradient centrifugation of dissociated sponge cells in order to separate the different cell categories according to their buoyant density. They then assayed the cell fractions for aggregation-promoting activity. Their initial results indicate that factor production is a function of the rather immobile secretory or 'mucoid' cells – for example, most of the cells with inclusions. This needs further investigation, since the cytoplasm in most of these cells is much reduced and their appearance is not that of cells engaged in protein synthesis. They may be storage phases only; however, they must be present for aggregation to proceed.

Another idea raised by Weinbaum and Burger (1973) and Turner and Burger (1973) is that aggregation factor, at least in *Microciona*, carries a terminal glucuronic acid residue which is necessary to allow the factor molecules to bind to specific recognition sites on cell membranes. This binding site has been termed 'baseplate'. On the basis of a considerable body of experimental evidence these workers have proposed protein–carbohydrate factor molecules linking to each other, and to membrane-located baseplate molecules, as a general model for cell-to-cell recognition.

This is a crucial and exciting area of current research with implications far beyond the sponge context. Certainly the answers to the outstanding questions, such as what is the active component of factor, where is factor located, what cells synthesize factor and when does it exert its effect during aggregation, will soon be available.

2.4 Morphogenesis and cell differentiation

There is considerable uncertainty about the stability of differentiated states in sponge cells and it has been suggested that some cell types can de-differentiate and embark on alternative developmental courses.

It is a fact that archaeocytes can give rise to all cell types of a mature sponge. The gemmules of fresh-water sponges before germination contain only specialized archaeocytes laden with food reserves. There is evidence from the study of early organization in aggregates that, in the process of reconstituting a pinacoderm, collencytes can become pinacocytes, Larval ciliated cells are said to become choanocytes at metamorphosis, and certainly later in development may become gametes.

There is little evidence for reversibility of these or any other cell transformation. Some differentiations, as for example from archaeocyte to a cell with inclusions, are known to be terminal; the cells either disintegrate and are phagocytosed or they are expelled. Spongoblasts and scleroblasts disintegrate after their skeleton-forming function is discharged, but the flexibility of, for example, a choanocyte or myocyte in terms of change of structure and function is hard to establish. The demonstration that choanocytes can become nucleolate and migrate from choanocyte chambers to become archaeocyte-like egg precursors (see Chapter 4) suggests that change of function and differentiation is possible, but only in one direction. Differentiation to an egg is irreversible; it is followed by fertilization or resorption and the cell cannot become a choanocyte again. Thus an adult sponge at all times must hold a population of archaeocytes that are not canalized but are available to embark on any line of development. This population must also sustain the major phagocytic and digestive functions for the organism. Repair of basal pinacoderm cells is slow following damage, as compared to exopinacoderm. This suggests that these cells cannot just be replaced by expansion of adjacent exopinacocytes, but that they represent a further and special differentiation from the collencyte population.

The few facts we have suggest a crucial role for three cell types – archaeocytes, choanocytes and collencytes. Some experiments by Borojevic (1966*b*) illustrate this and allow an assessment of the morphogenetic role and differentiative capacity of these cells in particular.

In the free-swimming larva of *Mycale contarenii* a substantial amount of cellular differentiation has taken place. There is a peripheral layer of ciliated cells broken only by non-ciliated pinacocytes at the posterior pole, a large posterior concentration of globoferous cells, and a central posterior aggregation of scleroblasts. The central region is predominantly archaeocyte and, except at the posterior pole, the subsurface layers are chiefly composed of collencytes (Fig. 2.6*a*).

This cellular stratification permits separation of different regions of the larva in such a way as to almost completely exclude certain cell types. The excised portions can be dissociated and their ability to form aggregates studied. Entire larvae, dissociated and allowed to aggregate, will produce a small sponge; they do not reconstitute larvae. These studies of the behaviour of portions of larvae, therefore, can give at least a guide to the behaviour of the different cell types during normal larval metamorphosis.

If the central mass of the larva is isolated (Fig, 2.6*b*) the cultures will contain predominantly archaeocytes with varying numbers of collencytes, scleroblasts and globoferous cells, but no ciliated cells. If the culture is very rich in archaeocytes and low in collencytes the aggregates will not flatten on the substrate and will never form a pinacoderm. Thus no mesohyl development can proceed and

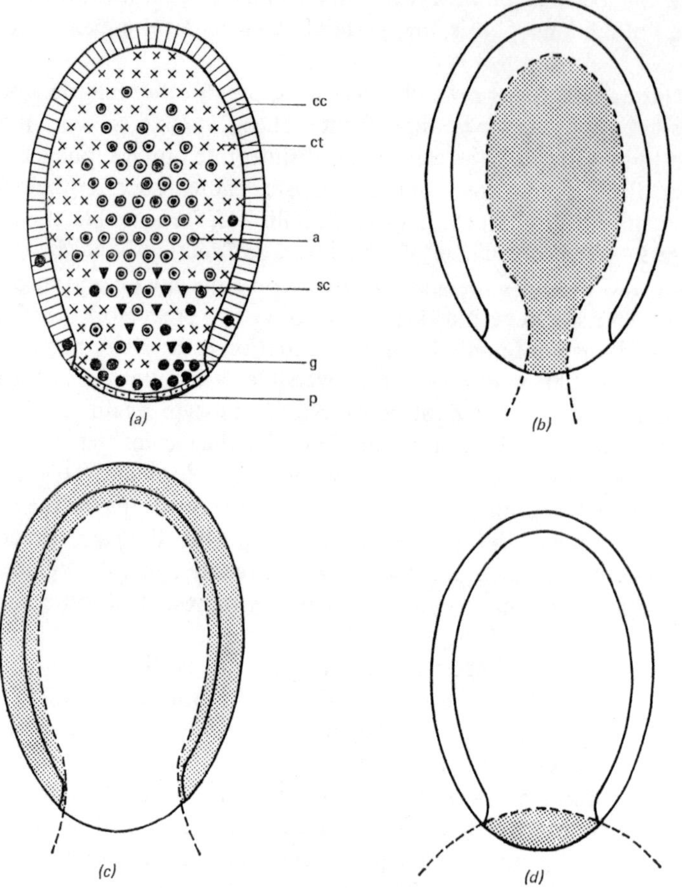

Fig. 2.6
(*a*) Diagram of the disposition of cell types in the parenchymella larva of *Mycale*. cc, ciliated cells; ct, collencytes; a, archaeocytes; g, globoferous cells; p, pinacocytes; sc, sclerocytes.
(*b*) In an experiment to determine the differentiation potential of the central region of the larva, the area cross-hatched was removed and cultured.
(*c*) A similar experiment on the superficial cells would involve the area cross-hatched. It is impossible to dissect columnar ciliated cells away without removing some underlying collencytes.
(*d*) Cultures of the posterior region only utilize the cross-hatched area, which contains only pinacocytes and globoferous cells.
(Redrawn from Borojevic 1966*b*.)

the aggregates dissociate and die after about ten to twelve days. If, however, the culture contains sufficient collencytes, the aggregates flatten, rapidly develop a basal and marginal pinacoderm and attach to the support within 24 h. The centre of these 'diamorphs' is composed of groups of archaeocytes in a dense fibrillar collagen matrix. Archaeocyte groups then differentiate progressively into choanoblasts and then to choanocytes until finally, after three to four days, choanocyte chambers are formed, inhalant and exhalant canal systems are established, and the sponge is functional.

Cultures of the posterior cap of the larva contain only pinacocytes and globoferous cells (Fig. 2.6*d*) and cannot develop. If some of the tissue adjacent to the posterior cap is included the cultures gain both collencytes and archaeocytes and develop to completion, as do cultures of the central region. Globoferous cells do not participate in morphogenesis and their role in the larva, as in the adult, remains unknown.

Cultures of the larval ciliated surface (Fig. 2.6*c*) will always contain a small percentage of archaeocytes and collencytes which are packed close to the base of the ciliated cells. If not dissociated, the ciliated surface will continue to swim, in disoriented fashion, after the central and posterior tissue has been removed. The larval shell settles by its anterior pole, a pinacoderm covers the ciliated surface to form a regular sphere which then flattens slowly on the surface. The central mass of the 'diamorph' remains undifferentiated and the subsurface area soon contains many large choanocyte chambers. Ciliated cells which are not incorporated in chambers are phagocytosed or cytolyse. The few archaeocytes present are always full of phagosomes and dividing actively. If the new culture survives long enough, the number of collencytes and archaeocytes is augmented to a normal balance in four days, and a functional sponge can be formed. Again, in this type of culture, if insufficient collencytes are included no attachment can take place and thus no development.

Several important generalizations can be derived from these observations. All cultures must pass through the same stage in their morphogenesis; the formation of a basal lamina and surface pinacoderm is indispensable and is a prerequisite for further morphogenesis. After this, differentiation of functional tissue proceeds from the edge of the 'diamorph' towards the centre, while the central mass remains a reserve of cellular material until the periphery is differentiated. This mode of development is common to all morphogenetic processes in sponges, whether in larvae after fixation, in reconstituting sponges from cell suspensions, or in development from gemmules.

The role of collencytes in these processes is fundamental. It is the percentage of collencytes in a culture which determines whether development will proceed; lack of choanocytes can be surmounted and if archaeocytes are sparse they can rapidly be replaced by division. A crucial minimum of collencytes must be present to initiate morphogenesis by securing and isolating the sponge. Once this has been achieved archaeocytes can complement the collencyte population for purposes of collagen secretion, and other differentiation can proceed.

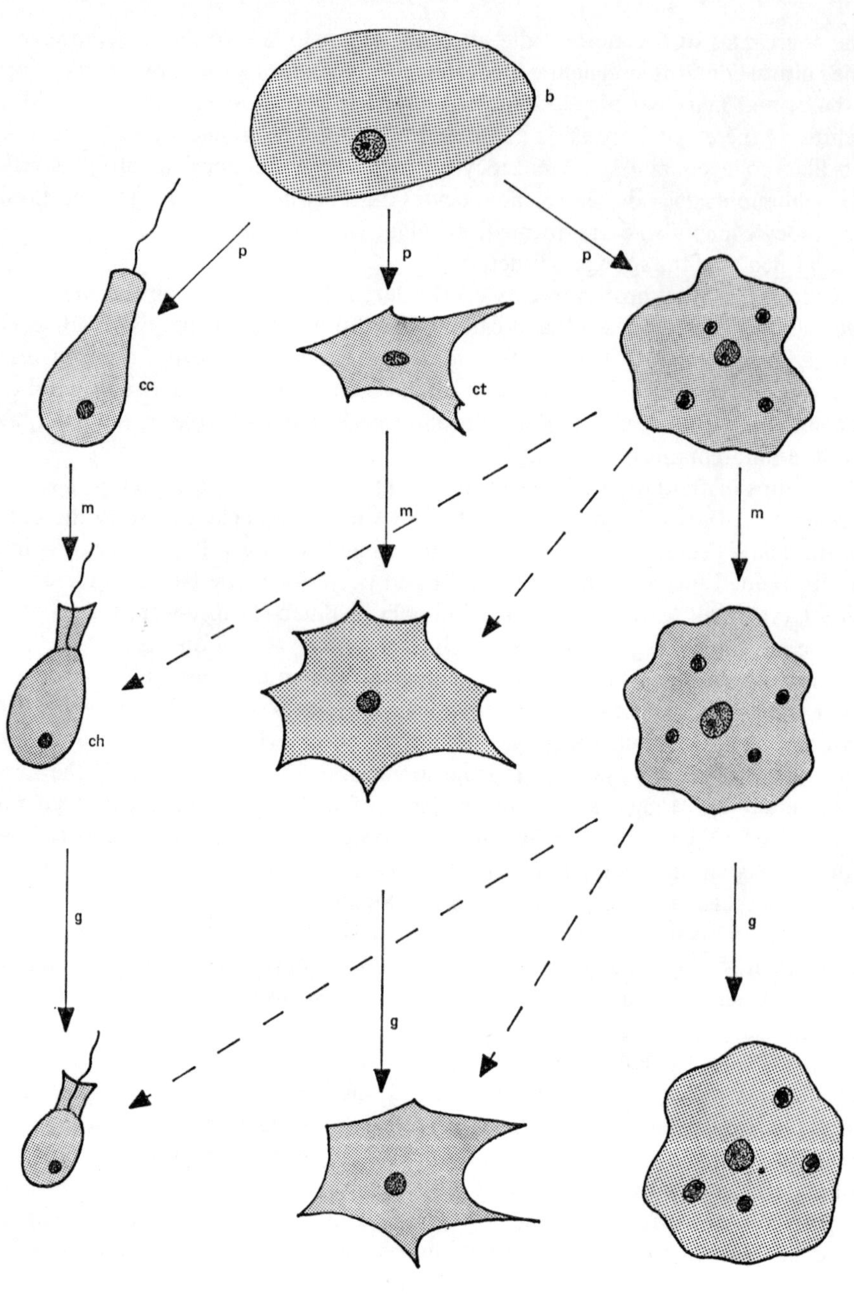
b
p
p
p
cc
ct
m
m
m
ch
g
g
g

Fig. 2.7 A schematic representation of the differentiation of the three principal cell categories which are essential for the early development of a sponge. (Redrawn from Borojevic 1966*b*.)

From the blastomere (b) the primary differentiation (p) achieved during larval life produces ciliated cells (cc), collencytes (ct) and archaeocytes. At metamorphosis (m), each cell type can retain its own identity with ciliated cells becoming choanocytes (ch). In addition archaeocytes can augment the population of both choanocytes and collencytes. During growth (g) the same situation persists.

Dashed arrows indicate differentiation of archaeocytes into other cell types. Solid arrows indicate the continuity of cell lines.

The fact that cultures of the posterior region of the larvae, which contain pinacocytes and globoferous cells only, cannot flatten or develop indicates that pinacocytes cannot replace collencytes in morphogenesis. In spite of the strong structural similarity of pinacocytes and collencytes, a physiological difference clearly exists between the two cell types. It seems that ciliated larval cells are transformed directly into adult choanocytes; those which are not are phagocytosed and cannot differentiate to other cell types. While this is true for reorganization following dissociation, it is open to question as a normal morphogenetic sequence.

The role of archaeocytes in a sponge is clear. On one hand they provide a reserve of totipotent cells, on the other they are the phagocytes. In this dual capacity they provide a regulatory mechanism establishing and maintaining the equilibrium between different cell types.

In view of these facts it is probable that differentiation in sponge cells is stable, i.e. that de-differentiation does not occur. Changes may take place in the morphology of particular cells in some situations, but function, based on the physiological and biochemical attributes of a cell, persists.

After the primary differentiation in development from a blastomere to either ciliated cell, collencyte or archaeocyte, the reservoir of archaeocytes remains to augment numbers of collencytes or choanocytes or to differentiate in other ways to establish the histological system characteristic of the adult sponge (Fig. 2.7). These additional, later differentiations, from the archaeocyte pool are irreversible, the material vested in these cells can only be reutilized by means of phagocytosis of expended cells by archaeocytes. Some capacity for further development is retained by collencytes which can become pinacocytes or myocytes and by choanocytes which can form gametes.

3 The sponge skeleton

3.1 Introduction

Reference has been made earlier to the components of the sponge skeleton and to the cells responsible for their synthesis. The skeleton is such an integral, unique and important a feature of sponges that it deserves consideration in further detail.

Porifera exhibit great diversity in their skeletal combinations and, because of this, the most fundamental divisions in their classification relate to skeletal composition and structure. The fact that skeletal components sometimes survive fossilization has resulted in current ideas on the evolution of sponges being guided by skeletal structure and organization. The intercellular matrix of the sponge mesohyl is in every way comparable to connective tissues of more complex organisms. Thus studying the synthesis of the primitive sponge connective tissue has much to contribute to the understanding of the biochemistry, synthesis and organization of such skeletal tissues in other Metazoa.

Two types of skeletal component occur in sponges: organic and inorganic. The former is always collagenous and the latter either siliceous or calcareous.

3.2 Collagenous structures

Collagen, the major structural protein of the animal kingdom, usually occurs as individual fibrils which vary little in structure and composition from group to group. In some cases, however, the precursor material to this collagen, procollagen, is organized differently into unusual structural arrangements which are peculiar to groups of animals, or to particular tissues.

In all attributes of their biochemistry studied so far, sponges have proved to be extremely diverse (see Chapter 7); thus it is not surprising that they display considerable polymorphism with respect to collagenous structures. Two major types of collagen occur: dispersed fibrillar collagen which is the same throughout the group, and spongin, which varies in structure and composition in special situations.

3.2.1 Fibrillar collagen

This is dispersed throughout the intercellular matrix (Pl. 1*b*) and is the only skeletal material which is found in all sponges. It shows the classic arrangement of collagen molecules polymerized from tropocollagen precursor units, being organized into fibrils ranging from 20 nm to 30 nm in diameter, with transverse banding at approximately 64 nm intervals (Fig. 3.1*a*, *b*, *c*). In sponges many cell types are capable of secreting fibrillar collagen; lophocytes and collencytes have this as their primary function.

It is now thought that wherever it occurs collagen is not finally organized within the cells which secrete the precursor material. However, in sponges there is some evidence that organization does sometimes occur inside vesicles in collencytes (Pl. 1*b*) as well as on the membrane at the point of secretion. In electron micrographs it is common to see collagen fibrils attached to and trailing from sponge cells (Pl. 2*a*). The process which involves the participation of membrane-bound elements may be the addition of sugars to the protein polymer, it may involve the polymerization process itself, or it may be that the enzyme responsible for polymerization is located on the membrane.

The extent to which fibrillar collagen reinforces the sponge extracellular matrix varies greatly within the phylum. It has been reported that no collagen is present in the trabecular network which constitutes the mesohyl of Hexactinellida. The only relevant modern study (Travis, François, Bonar and Glimcher 1967) demonstrates that collagen occurs in traces in the organic framework of the siliceous skeleton. It is difficult to envisage how a cellular reticulum can stretch around a complex spicule network without some organic structural support, and this 'peculiarity' of hexactinellid sponges has again proved illusory. The unpublished observations of Bergquist and Nelson on desilicified, Mallory triple-stained sections of *Rossella nuda* demonstrate that a thin collagenous sheath surrounds each hole from which a spicule has been dissolved. This sheath is seen in electron micrographs to be an aggregation of fibrillar collagen elements. Fibrillar collagen in the trabecular region is therefore present, but sparse, in the Hexactinellida. In Calcarea also, the only collagen present is fibrillar, and dispersed lightly through the mesohyl, never occurring in dense concentrations. In Demospongiae fibrillar collagen is universal and obvious, often forming very dense bands as in the cortex of *Tethya*. In two groups of Demospongiae, the Homosclerophorida and the Dendroceratida, some genera (e.g. *Oscarella, Halisarca, Bajalus*) have only fibrillar collagen as skeletal support material, the sponge body in all cases being thin and fragile while cellular mesohyl elements are much reduced.

Traditionally the organic sponge skeleton has been referred to as a 'spongin' skeleton. Since the identification of fibrillar collagen as a major component of the sponge matrix, and the discovery that it differed distinctly in structure and chemical composition from the more obvious fibrous elements, two terms have been used: spongin A for the fibrillar intercellular collagen and spongin B for the large horny fibres and related structures. In effect, the term spongin implies that the substance is unique to sponges and this is true of spongin B type structures. However, numerous studies on the molecular organization and composition of collagen confirm that fibrillar dispersed collagen is the same

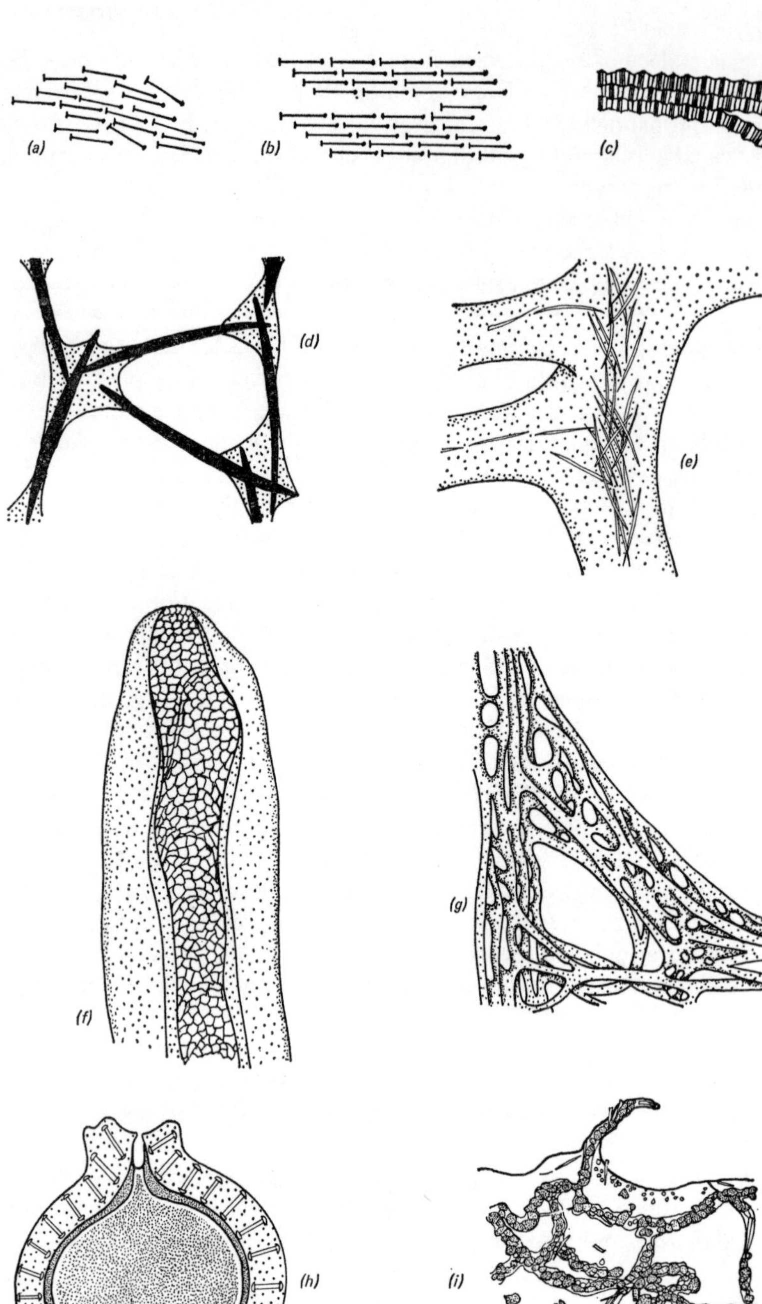
(a)
(b)
(c)
(d)
(e)
(f)
(g)
(h)
(i)

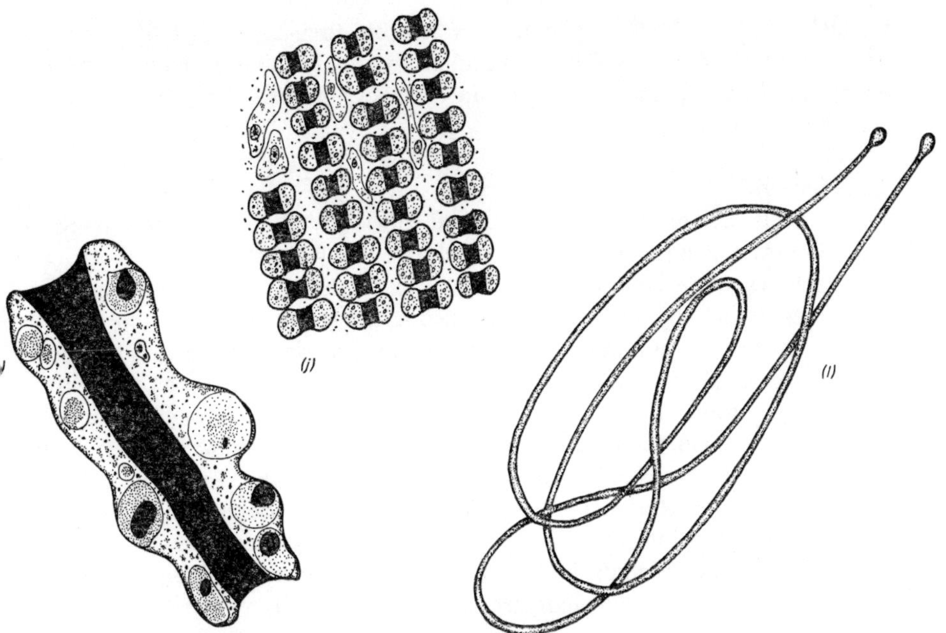

Fig. 3.1 Collagenous skeletal elements of sponges.
(*a*) Tropocollagen molecules, the precursor units of collagen.
(*b*) With orientation, tropocollagen units form collagen.
(*c*) Once the units are aligned to form collagen fibrils a transverse banding, or periodicity, is apparent.
(*d*) Perispicular spongin, where patches of spongin cement spicules at points where groups intersect, as for example in some species of *Haliclona*.
(*e*) Perispicular spongin where the spongin forms a major fibre which has a spicule core. This arrangement is found in many Haplosclerida.
(*f*) A spongin fibre of *Verongia* in longitudinal section. In this genus there is an outer dense 'bark' and an inner, less compact, pith region in all fibres.
(Redrawn from Vacelet 1971*b*.)
(*g*) A portion of the complex homogeneous spongin fibre skeleton of *Sarcotragus*.
(*h*) A fresh-water sponge gemmule which has an outer spongin coat in which amphidisc microscleres are embedded.
(Redrawn from Evans 1901.)
(*i*) A spongin fibre skeleton in which sand grains and other fine debris have been incorporated. This type of construction can become very elaborate.
(*j*) Diagrammatic representation of a tract of intracellular segmented fibres, as seen in *Haliclona*.
(Redrawn from Meewis 1948.)
(*k*) A single cell of a tract of segmented fibres. The large vesicular inclusions which remain in the cell after the central tract is formed resemble the vesicles of spherulous cells in *Haliclona*.
(Redrawn from Levi 1967.)
(*l*) A spongin filament as found in the genera *Ircinia* and *Sarcotragus*.

throughout the animal kingdom. It is sensible then to restrict 'spongin' to collagenous structures which are peculiar to sponges and to equate fibrillar collagen with comparable structures in other organisms. The term spongin A should not be further used.

3.2.2 Spongin

The multiple uses of spongin as a skeletal substance within the Demospongiae are only now becoming obvious and the differences which exist between various spongin types raise serious questions as to whether the same spongocytes are able to synthesize and secrete two or more distinct chemical substances. At present the details of synthesis and fine structure are not complete, but there is sufficient information to allow five different skeletal structures to be recognized within the spongin group. These are spongin fibres and spicules, spongin filaments, perispicular spongin, the spongin case of gemmules and the strange small spongin fibres formed by chains of specialized cells in *Haliclona* and *Adocia*.

All of these structures share one basic attribute. Although fibrillar, their molecular organization is not as rigid as that in true fibrillar collagen, and thus periodic transverse banding is discernible only occasionally. Spongin microfibrils are around 10 nm in diameter, and the periodicity, when developed, is around 60 nm. Differences in X-ray spectral characteristics, amino acid and sugar composition exist between these substances in all cases where analysis has been attempted.

(a) Spongin fibres

These form complex networks in many sponges and frequently attain a thickness of several millimetres. They often incorporate siliceous spicules (Fig. 3.1*e*) or sand grains (Fig. 3.1*i*).

Some genera of the Dictyoceratida and Haplosclerida have a very characteristic and complex organization of the spongin fibre skeleton and a size hierarchy of primary, secondary and tertiary fibres is recognizable (Pl. 5*b*). In *Spongia*, for example, the large primary fibres contain a central core of sand and spicule detritus, but the more numerous secondaries are clear of inclusions. In *Hippospongia* the secondary fibre network is emphasized greatly at the expense of the primaries and thus almost no foreign material is incorporated into the fibre network. Hence the prime quality of some *Hippospongia* skeletons for bath sponges. Many people conditioned to the idea of bath sponges are surprised to learn that this flexible, tough, familiar object is only the skeleton of a very fleshy organism. A spongin fibre skeleton confers great strength on the living sponge, but affords no deterrent to predation; horny sponges are very frequently eaten by fishes, opsithobranchs and echinoderms.

Microscopic observation of spongin fibres shows concentric growth lines in all cases, and frequently spongocytes are seen opposed closely to the surface of the fibre, releasing material in globules onto the growing structure. In most cases, spongin fibres are homogeneous in cross-section, but in some families,

Verongiidae and Aplysillidae for example, there is a dense outer bark of compacted spongin surrounding a pith region of more open composition (Fig. 3.1*f*). At growing points in such fibres where no bark has been secreted, the secretion of pith by spongocytes has been observed. It appears that spongocytes crowd to the growing tip and secrete material to form the pith and then elongate and slide slowly down the fibre, laying down bark as they move. This behaviour could account for the concentric nature of the dense outer fibre (Fig. 3.2).

In one genus of Dendroceratida, *Darwinella*, spongin spicules are produced, these are diactinal, triactinal, or tetractinal and dispersed through the sponge. Spongin fibres of typical form are also present.

(b) Perispicular spongin

The skeleton of many Demospongiae is composed of siliceous spicules cemented into a regular arrangement by spongin deposited at their points of intersection (Fig. 3.1*d*), or deposited in such a way as to enclose single or multiple tracts of spicules (Fig. 3.1*e*). Recent studies on the composition and secretion of this perispicular spongin (Garrone and Pottu 1973) show that it resembles closely

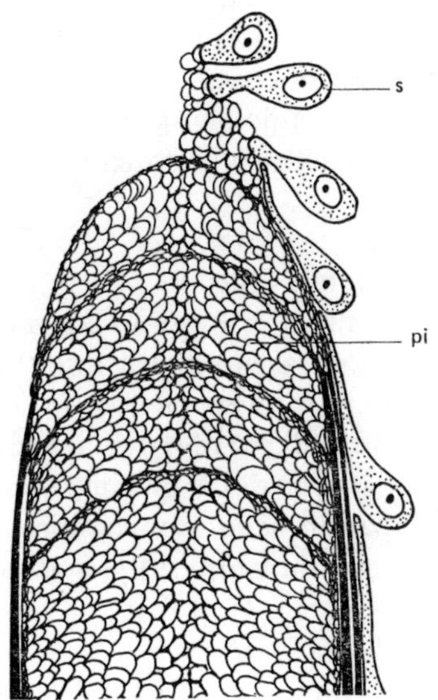

Fig. 3.2 The method of fibre secretion in *Verongia*.
Spongocytes (s) congregate at the growing apex of the fibre and lay down components of the pith (pi). They then elongate and, still adhering to the fibre, glide away from the apex. Successive waves of spongocytes thus form the concentric, dense outer bark.
(Redrawn from Vacelet 1971*b*.)

the spongin which makes up the pith in spongin fibres (Fig. 3.1*f*). In young growing sponges, the positioning and securing of the spicules in the correct spatial relationship is a precise morphogenetic process which requires organized, accurate collagen synthesis by groups of spongocytes acting together (Pl. 1*a*). Combination of spongin and spicules allows the development of very flexible, strong skeletal structures, which, since they tend to be similar in large groups of sponges, can be most useful in classification.

(c) Spongin filaments

Two genera of horny sponges, *Ircinia* and *Sarcotragus*, contain, in addition to collagen fibrils and spongin fibres, peculiar filaments dispersed through the mesohyl. These are several millimetres long, up to 10 μm wide, and terminate at each end, in a knob (Fig. 3.1*l*). In transverse section the filaments are three layered, with an outer, carbohydrate-containing cuticle, a central axial thread and between these, a cylinder of uniform material. At high magnification the whole structure is revealed as bundles of microfibrils, 5–7nm in diameter, twisted in helical fashion, nearly straight in the axis, and increasingly coiled toward the periphery. Physical and chemical analyses show these filaments to be collagenous and very similar to spongin fibres.

Secretion of filaments is by nucleolate cells which liberate amorphous material to the matrix where it is organized to produce the regular collagen pattern. In spite of the similarity in composition of filaments and spongin fibres, it is not surprising that the mode of secretion and growth is quite different. The disposition in the sponge and the macroscopic structure of the two collagen types is very distinct.

(d) Spongin forming the gemmule coat

The coat of the asexual gemmules of fresh-water and some marine sponges is another example of a special collagenous structure (Fig. 3.1*h*). It provides the desiccation resistance necessary to overwintering in severe climates. In general structure and composition, the collagen of the gemmule coat is extremely close to that of spongin fibres, but it includes individual giant collagen fibrils which are up to ten times larger than those in spongin fibres, and in addition, a large quantity of amorphous material in which no periodicity can be discerned.

(e) Intracellular segmented fibres

Several sponges of the order Haplosclerida can develop long chains of precisely oriented cylindrical cells, each containing a central rod of spongin-like material (Fig. 3.1*j*, *k*). The collagenous nature of this material has not been confirmed unequivocally. These strings of cellular segments do, however, function as skeletal elements and act essentially as tiny spongin fibres. In the Haplosclerida, where growth of the sponge is rapid and changes in body shape are frequent, such a simple system of supple fibres could provide necessary support during the formation of the more rigid spicule and spongin skeleton.

3.3 Mineral skeleton

A mineral skeleton, either of calcium carbonate or silica, is found in all sponges except those belonging to the Dictyoceratida, Dendroceratida, Verongida, two genera of Homosclerophorida, and one genus of the Hadromerida.

The most common type of inorganic skeleton is the spicule skeleton, where individual cells, singly or in groups of up to four, form discrete spicules. These, according to their morphology and disposition in the sponge, are termed either microscleres or megascleres. This terminology is applicable to Hexactinellida and Demospongiae; Calcarea do not have microscleres. In some groups, discrete spicules are either supplemented or replaced by more massive structures. An ancylosed spicule skeleton is found in all Lithistida, and massive calcitic and aragonitic skeletons occur in Pharetronida and Sclerospongiae, respectively.

There is nothing more boring than descriptions of the diversity of spicule forms in sponges, and in the final analysis nothing so misleading as a classification which leans too heavily upon such descriptions. Spicules can be described precisely, and it is easy to forget that although other aspects of sponge structure are less amenable to measurement and description, they can be just as informative in a phyletic sense. It is, however, necessary to introduce a few terms to allow discussion of sponge spicules. Siliceous spicules have an axis of organic material around which the inorganic component is deposited. Some disagreement is found with respect to calcareous spicules, where the only trace of axial structure is an optically evident axial region.

Descriptive terms which designate the number of axes end in '-axon'; those designating the number of rays end in '-actine' or '-actinal'. In addition to this there is an elaborate nomenclature which specifies the shape, size and ornamentation of a given spicule.

It is best to take a few examples. Perhaps the commonest spicule type is the oxea; this is smooth, tapering to two similar and pointed ends. It is a diactinal monaxon or a diactine (Fig. 3.3*a*). The tylostyle, typical megasclere of the order Hadromerida, is knobbed at one end, pointed at the other. It is a monactinal monaxon (Fig. 3.3*e*). During growth of the spicule, if development proceeds from a central point in two opposite directions a diactine will result. If growth proceeds from a fixed point in one direction only, a monactine is produced. Growth in four directions, equally and symmetrically around a central point, produces the tetractinal calthrops spicule typical of the Homosclerophorida (Fig. 3.3*i*; Pl. 6*b*).

Further variation occurs with the addition of spines to the spicule surface (Fig. 3.3*g*, *h*), a condition denoted by the prefix 'acantho'; with the multiplication of rays in microscleres to produce polyactines, as for example in the whole range of star-like, or asterose, microscleres (Fig. 3.4*g*, *h*, *i*,; Pl. 6*a*); and lastly by the unequal development of rays, as in the triaene spicules of the Choristida, where normally four rays make the tetraxon pattern, but three, the cladi, are reduced and one, the rhabdome, is accentuated (Fig. 3.4*f*, *j*, *k*, *l*). Hexactinellida

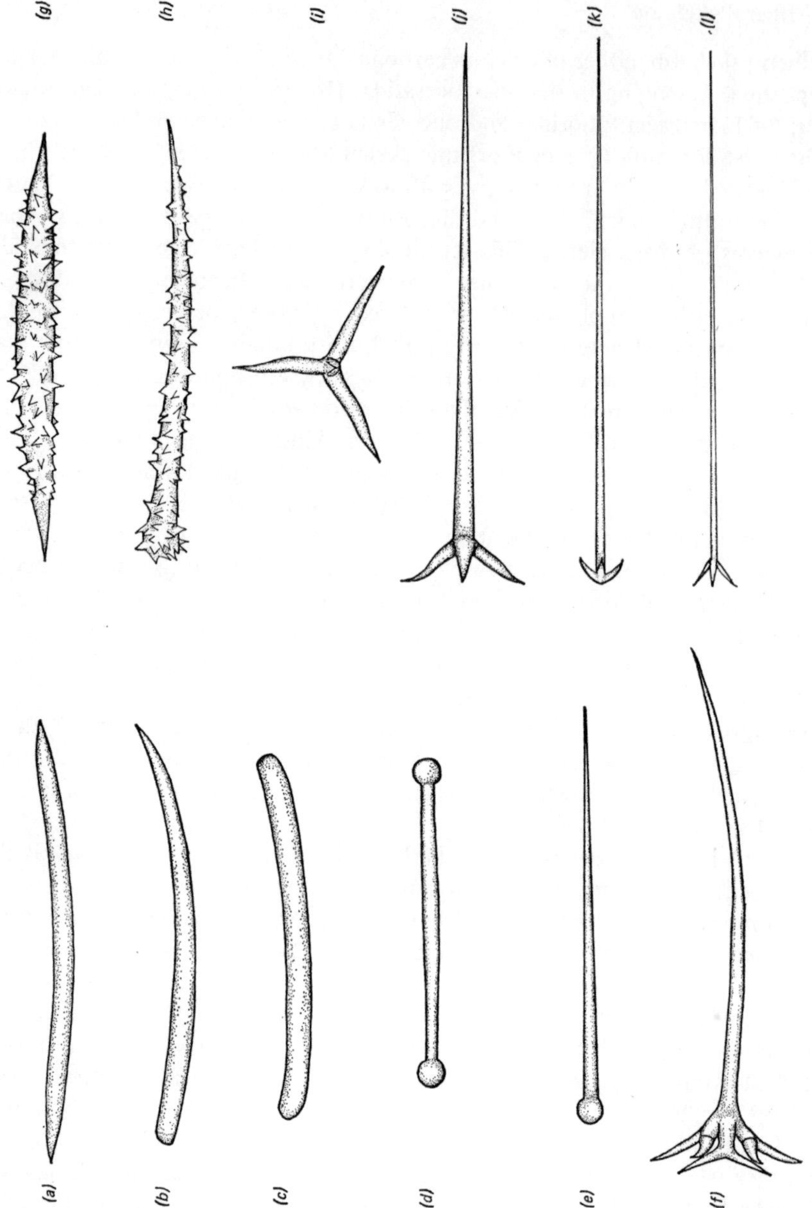

Fig. 3.3 Representative megasclere types of the Demospongiae.
(*a*) oxea; (*b*) style; (*c*) strongyle; (*d*) tylote; (*e*) tylostyle; (*f*) plagiotriaene; (*g*) acanthoxea; (*h*) acanthostyle; (*i*) calthrops; (*j*) orthotriaene; (*k*) anatriaene; (*l*) protriaene.

have, in addition to other spicule forms, a triaxon or hexactinal configuration (Fig. 1.6*c*) and in the Calcarea a triaxon but triactinal spicule is common.

The form of microscleres is extremely variable and cannot easily be summarized. There are three major types: asterose forms, where rays diverge from a central point; sigmoid and chelate forms, where a single axis is twisted or recurved and ornamented; and raphides, rhabds or toxiform types with a single axis which is either straight or flexed in one plane. Variants upon these themes occur and in some cases these characterize orders and families both in Hexactinellida, and in the Demospongiae, where the greatest diversity of microsclere forms, and indeed of all skeletal elements, is to be found.

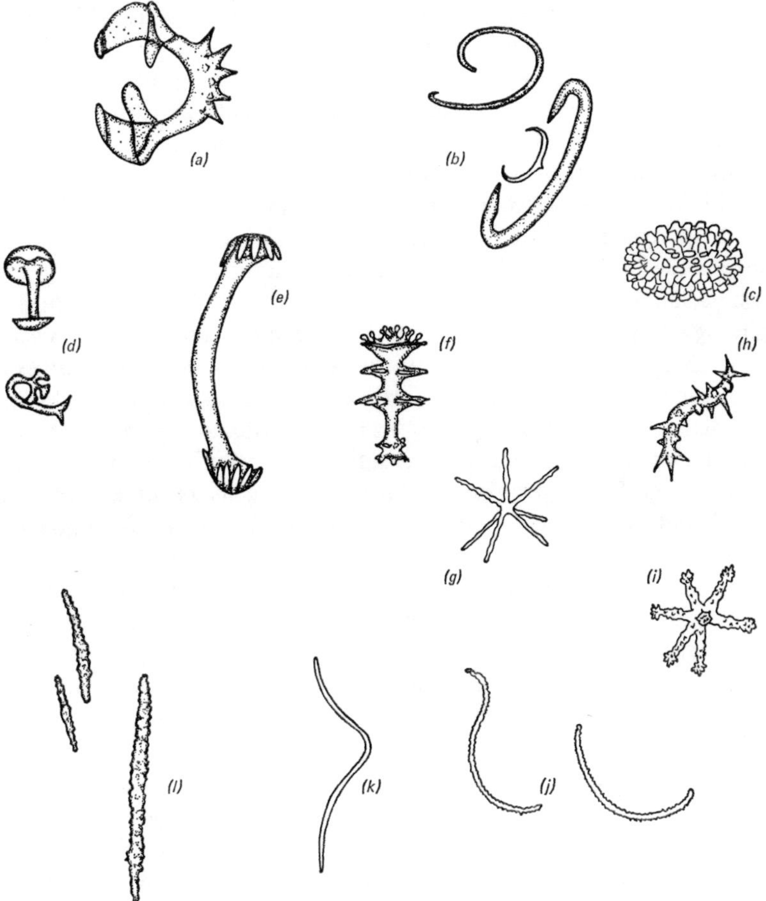

Fig. 3.4 Representative microsclere types of the Demospongiae.
(*a*) An anchorate isochela with ornamented shaft; (*b*) a group of sigmas; (*c*) a sterraster; (*d*) bipocilli; (*e*) an arcuate isochela; (*f*) discorhabd; (*g*) euaster; (*h*) spiraster; (*i*) acanthotylaster; (*j*) sigmaspirae; (*k*) toxa; (*l*) acanthose microxeas.

One major classification of the Demospongiae has emphasized microsclere form, either sigmoid or asterose, as a character to define two subclasses, Astrotetraxonida and Sigmatotetraxonida. It is, however, clear that the evolution of microscleres has been subject to few constraints, and that any form will serve the simple function of reinforcing membranes. Sigmoid forms (Fig. 3.5*b*, *j*) in particular have evolved in widely separate lines of Demospongiae; thus classification on this basis is highly artificial.

3.3.1 Functions of the mineral skeleton

A spicule or solid skeleton is primarily a supplementary supporting structure. It can act to confer a slight degree of rigidity to a spongin fibre skeleton and, if the amount of inorganic material in relation to organic material is increased, the sponge becomes more and more solid until the texture, in some Choristida and Lithistida, is like rock. The support function of a spicule skeleton can be supplanted by fragments of sand, rock or fragments of extraneous spicules taken up by the sponge (Fig. 3.1*i*). When these 'foreign' skeletons are bound by spongin fibre they can be just as organized as endogenous spicule skeletons.

There is no evidence to support the view that a strong spicule skeleton renders a sponge less prone to predation by grazing organisms. In the case of the huge hexactinellids like *Rossella*, which have a high ratio of silica to organic material, echinoderms graze them with impunity and are able to alter the surface morphology substantially. Encrusting Demospongiae such as *Microciona* and *Haliclona* are grazed by opisthobranchs, often very selectively; for example *Rostanga rubicunda* which grazes on *Microciona coccinea*. In tropical and subtropical seas, grazing of sponges by fish and turtles is common. It is more likely that sponge defences against predation, if they exist, are biochemical not physical. Many Demospongiae produce toxic mucus or huge quantities of terpenoid and sulphur compounds with unpleasant taste and known toxic effects.

Thus, the main function of a megasclere skeleton is the maintenance of the gross form of the sponge and, in combination with spongin fibre skeletons, extremely large sponges with well-defined morphology can be supported. In such large sponges, inhalant and exhalant canals can be several centimetres in diameter and may need to traverse some thickness of spicule cortex en route to or from the choanocyte chamber region. The lining layer of these canals is of course made up of endopinacocytes and strengthening of the canal surfaces and of surface membranes is the function of microscleres. These spicules rarely form organized and oriented layers, as do many megascleres; they are effectively just little units of silica which act as packing between large spicules, or as reinforcing for delicate surfaces.

3.3.2 Secretion of spicules

Megasclere secretion has been well studied only in Calcarea. Only sketchy observations are available on the early growth of siliceous megascleres. Secretion

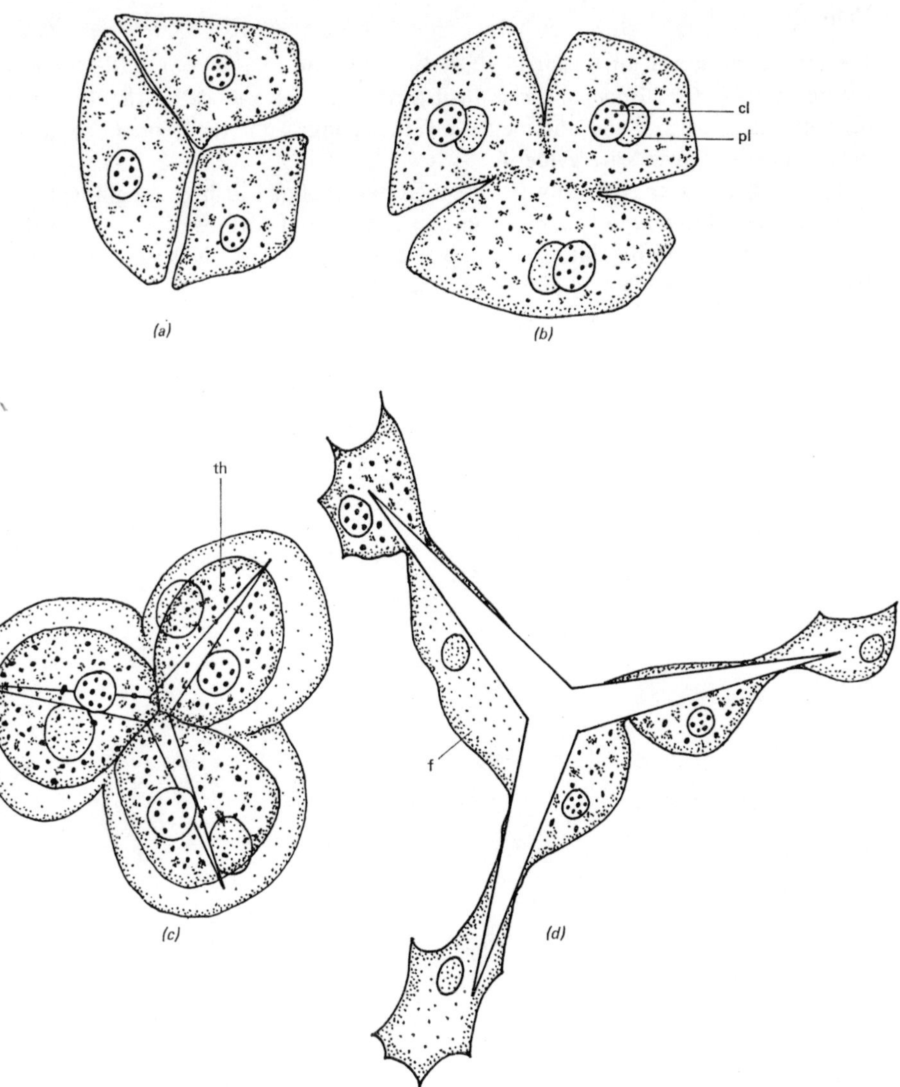

Fig. 3.5 The secretion of a calcereous spicule.
(*a*) Sclerocytes associating to form a triad.
(*b*) Nuclear division in each cell to produce central (cl) and peripheral (pl) nuclei.
(*c*) A calcitic ray is secreted between each pair of nuclei, and the inner group of three sclerocytes, the thickener cells (th), move outwards along the rays.
(*d*) The outer group of three cells, the founder cells (f), remain at the junction of the spicule axes for a time and then also migrate along the rays.
(Redrawn from Jones 1970.)

Plate 3

Electron micrograph of the myocyte network found in the contractile oscular membrane of *Hippospongia*. The fusiform shape of the cells, the oriented microtubules (mt) and microfilaments (mf) they contain and the frequent contacts between cells (arrows) are evident.

The insets at lower right show the aligned microtubules at high magnification.

Networks of this type between contractile cells and surface cells provide the structural basis for the integrative system as represented in sponges. (After Pavans de Ceccatty, Thiney and Garronne 1970.)

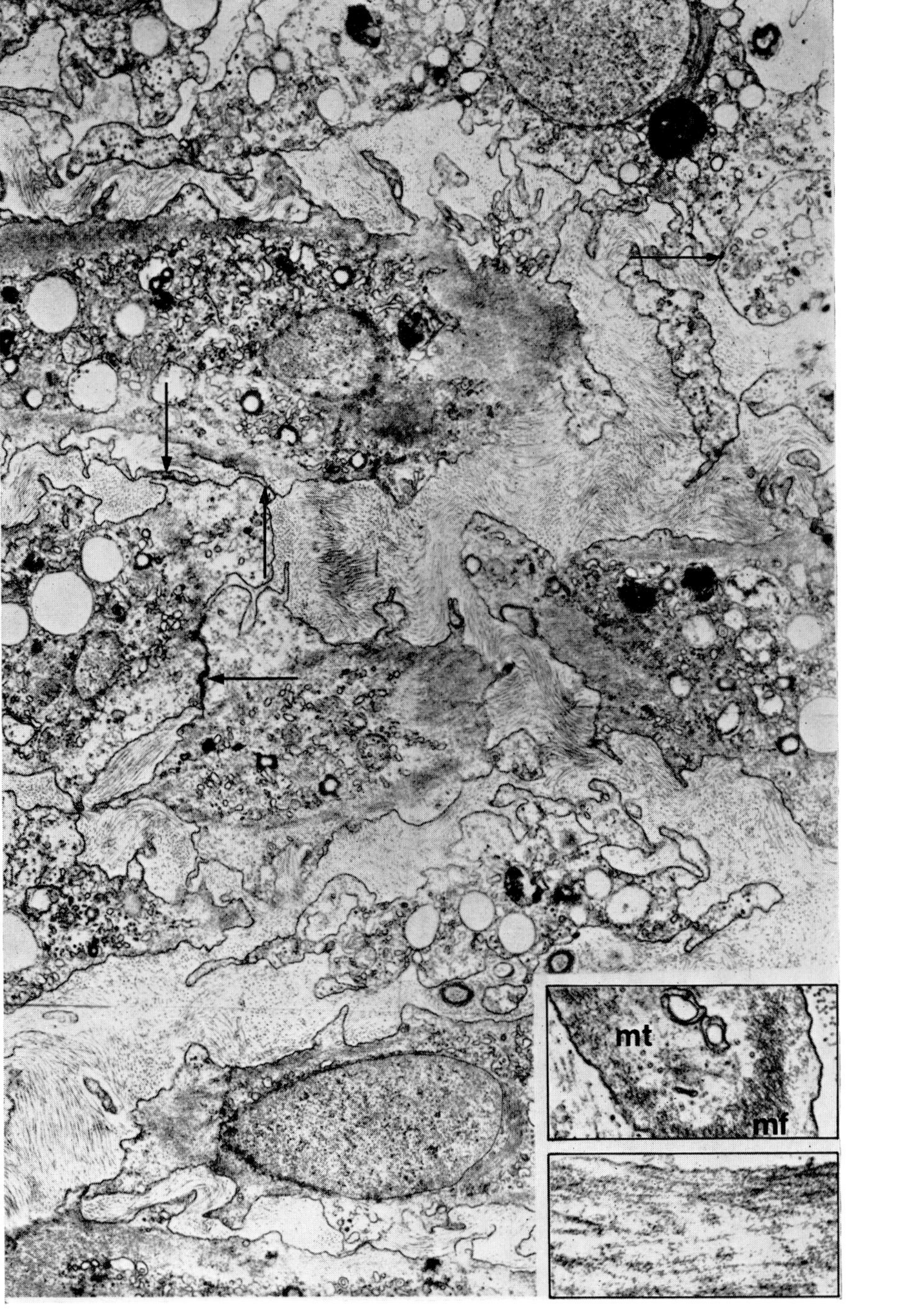
mt
mf

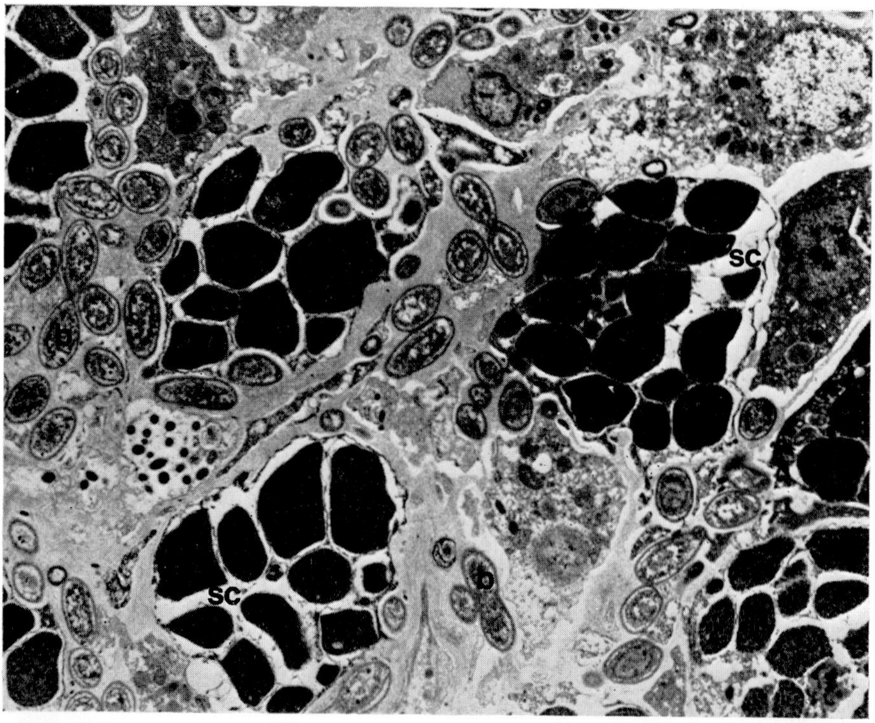

sc
sc
a

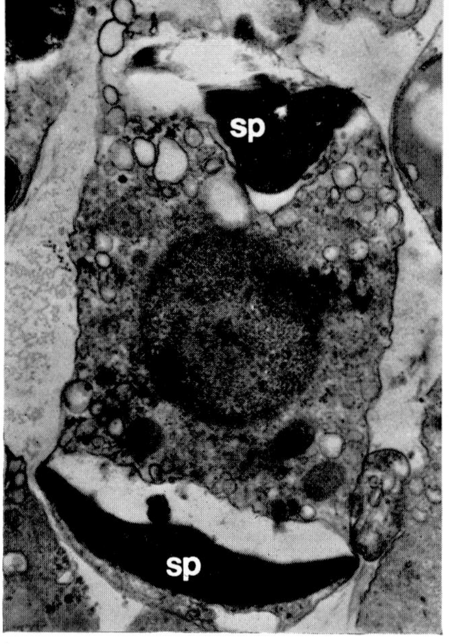

b

c

Plate 4

(a) An electron micrograph of a section taken just below the thick collagenous cortex of *Psammaplysilla*. In this region there are dense concentrations of spherulous cells (sc) and abundant extracellular bacteria (b), some of which can be seen dividing.

(b) A sclerocyte in the process of secreting two siliceous spicules (sp), one at each side of the cell inside a vacuole.

(c) Oocyte release in *Agelas*, an oviparous sponge. The individual in the foreground is covered by cords of the yellow mucus which surrounds the oocytes during at least their early development. Gamete release is clearly an eruptive event in this species, but appears not necessarily to involve all individuals in the population. The two specimens in the centre show no sign of oocytes. However, they could be male.

The photo was taken at 80 m depth, at Grand Cayman Island. (Reproduced from Reiswig 1976.)

in all cases is effected by sclerocytes, special cells which have the capacity to accumulate calcium or silicate and to deposit this in an organized way. In Demospongiae the silica is deposited on an organic filament which is hexagonal in section and composed of protein crystallized in three primary rays with 120° angle between planes of the ray. No organic axial template is found in calcareous spicules, but there is an optically evident 'axial region'.

Take the example of a triradiate calcareous spicule: these are secreted by sclerocytes, working two per spicule ray. The cells derive from the exopinacoderm and are recognizable by their coarse granular cytoplasm. Sclerocytes first associate to form a triad and then each cell divides in such a way that three nuclei are central and three peripheral (Fig. 3.5*a*, *b*). It is not absolutely clear that cytoplasmic separation is total in the early stages, but a ray of calcite is secreted between each pair of nuclei. The inner sclerocyte of each pair then moves progressively outwards from the centre, lengthening the ray as it goes. The outer sclerocytes remain at the junction, usually in the angles between the rays. The three outer cells were termed founder cells because they were responsible for lengthening of rays, while the outer three were called thickener cells. The latter remain at the central junction for a time, but eventually migrate along the rays, apparently thickening them as they go (Fig. 3.5*c*, *d*). Both cells eventually slide off the spicule ray. There are slight variations on this procedure for four-rayed spicules, where the apical ray is added in a slightly different manner, and for monaxons which can be formed in pinacoderm cells which are still *in situ*, but which behave as sclerocytes.

Calcareous spicules grow at the rate of 1.0–2.5 μm h^{-1}, and during growth, developing rays are enclosed by a membrane which must function as a calcium pump. The process producing the spicule calcite is clearly selective. The magnesium ion molar percentage of spicules of *Leucandra nivea* is 11.2 (Jones and Jenkins 1970). This could crystallize from a solution containing magnesium and calcium in a molar ratio of 4:1; sea water is about 5:1. The difference indicates that a selective, species-variable process operates in ion accumulation.

There are no comparable data available for siliceous spicule production in marine sponges, but simple diactinal oxeas and microscleres are known to be produced inside single mononucleate cells. In fresh-water sponges, Elvin (1971) has provided evidence for the operation of two distinct processes in spicule growth. The first of these is organic synthesis, involving axial filament deposition and organization. This governs spicule shape and has a major effect on final spicule length. The process slows as the spicule gets longer, is accelerated by increasing temperatures and is retarded by high silicic acid concentrations. Secondly, there is the process of silica deposition, the rate of which is governed by available spicule surface and silicic acid concentration. Typical deposition rates are $10\text{–}20 \times 1^{-18}$ mol SiO_2 μm^{-2} spicule surface min^{-1}. The actual transport of silicon from the matrix to the site of deposition along the axial filament involves first the scleroblast membrane and then the unit membrane enclosing the axial filament, the silicalemma.

4 Reproduction and development

4.1 Introduction

All sponges which have been studied in detail are known to be capable of sexual reproduction, although the mechanisms vary considerably from group to group. In addition, several types of asexual reproduction are common within the phylum. These range from simple regeneration following fragmentation, to production of internal and external buds, or formation of complex gemmules. It has also been reported that asexual processes can lead to the production of free-swimming larvae, which normally are the end product of a sexual process.

All regulatory aspects of reproduction are poorly known. Topics such as how fertilization occurs, what initiates spermatogenesis and oogenesis, what initiates gamete release, how long the reproductive period lasts, and what adult cells give rise to the gametes are poorly documented in most groups. There are several reasons for this deficiency. Great variability in almost all but the most general aspects of reproduction makes it impossible to predict detailed patterns of reproductive behaviour in any group of sponges which has not been studied specifically. The lack of any organized gonads or gonoducts means that there is no clear target area for study, as the gametes are dispersed throughout large areas of the sponge mesohyl. Then, finally, in a majority of the sponges there is a marked asynchrony within individuals and within populations in the reproductive process. Crucial events such as fertilization, early cleavages, and incorporation of parental material into developing embryos may be occurring in only a small number of embryos in a given specimen at the time of fixation. Chances of observing whole sequences are very low unless a whole sponge population can be studied intensively over a long period.

4.2 Sexual reproduction and embryonic development

For purposes of later discussion it will help to outline a generalized reproductive sequence and then point to some of the interesting departures to be found in particular groups. Demospongiae of the orders Halichondria, Poecilosclerida, Haplosclerida, Dictyoceratida and Dendroceratida have been frequently studied and, since this group of orders constitutes a cohesive unit within the subclass Ceractinomorpha, it is possible to give at least a general account of reproduction for this group from observations made on different genera and species by different workers.

4.2.1 A reproductive sequence in a viviparous sponge

In general, sponges are hermaphrodite, but produce oocytes and spermatocytes at different times. This successive or temporal hermaphroditism is difficult to distinguish from gonochorism unless the periods of female and male gamete production overlap. In the case of *Halichondria moorei*, a common intertidal sponge in New Zealand waters, there is a sexually dimorphic endosomal colour difference which allows male and female individuals to be distinguished during the reproductive period.

Gamete production and embryo development are not localized except that they are confined to the endosome where a major portion of the tissue may be involved at any time. It is usual to find normal mesohyl and choanocyte chambers interspersed with reproductive elements at different stages of development. In certain species, particularly those which are thin and spreading, there is a tendency for reproductive elements to occupy the base of the sponge, and in these situations the canals and choanocyte chambers are displaced toward the surface, or suffer some degree of disorganization.

The cellular derivation of male and female gametes has proved difficult to determine with certainty. There is good evidence, based on electron microscope examination of changes in cell morphology, that spermatocytes in *Aplysilla rosea* are formed by the transformation of choanocytes into primary spermatocytes (Tuzet, Garrone and Pavans de Ceccatty 1970). In all other groups, spermatic cysts have the dispersion, size, and general appearance which suggests they have developed by the transformation of choanocytes into sperm cells. Only Diaz, Connes and Paris (1973) have followed the transformation of choanocytes into oocytes. Using electron microscopy, they followed the addition of a nucleolus, changes in cytoplasmic inclusions and loss of flagellum and collar which are necessary stages in the process. The transformed cells then moved out of the choanoderm into the mesohyl. Apart from this careful observation on *Suberites massa*, an oviparous species, the general assumption has been that oocytes in Demospongiae arise from archaeocytes. This assumption is based on the position of the oocytes at the earliest recognizable stage before the cells assume a regular rounded shape. For the Calcarea, a majority of authors agree that oocytes are formed from cells of the choanocyte layer. It is not pertinent to fall back on the obvious fact that in certain circumstances, such as gemmule germination and culture of segments of larvae, archaeocytes are capable of forming choanocytes. We are concerned here with what happens in the normal course of reproductive events.

Spermatogenesis occurs within distinct spermatic cysts which can be formed when all the cells of a choanocyte chamber are transformed to spermatogonia. These are densely packed and separated from the mesohyl by a single layer of follicle cells. If cells migrate from the choanocyte chamber after transformation they can aggregate at points in the mesohyl to form small spermatic cysts. These aggregates, isolated by follicle cells, resemble spermatic cysts formed in choanocyte chambers. In some cases there are several spermatogonial generations, while in others choanocytes transform to primary spermatocytes directly. Division of the latter gives secondary spermatocytes. These then divide forming spermatids which produce sperm. The shape and relative development of the various regions of the sperm are variable. The length of the entire spermatozoon is between 10 and 15 μm. Development of sperm can be synchronous throughout the individual, synchronous within a cyst but variable throughout the sponge, or even consecutive within a spermatic cyst.

Oogenesis in the early stages proceeds without the apposition of follicle cells to enclose the oocyte which may be irregular in form, and possibly still mobile. At this stage oocytes are marked by their prominent nuclei (Fig. 4.1) and will differ in size from species to species. In *Haliclona ecbasis* (Haplosclerida) they are from 9 to 18 μm (Fell 1969) (Fig. 4.2*a*). When they reach a diameter of 20 μm in *H. ecbasis* ovocytes become ovoid, and are enclosed within a layer of flattened follicle cells (Fig. 4.2*b*). These cells, in electron micrographs, resemble endopinacocytes.

Increase in size of the oocyte in the early stages is not accompanied by accumulation of yolk but, in *H. ecbasis,* when the cell is around 30 μm in diameter nurse cells, whose role is to provide material for further growth, aggregate around the layer of follicle cells (Fig. 4.2*c*). The nurse cells are then engulfed and the oocyte diameter increases to around 140 μm (Fig. 4.2*d*, *e*). Most nurse cells loose their nuclei after they are incorporated into the growing oocyte, but their cytoplasm remains little changed. As this process proceeds the oocyte cytoplasm becomes so attenuated that it is difficult to see, and in sections these large oocytes appear as spherical masses of nurse cells (Fig. 4.2*f*). Fell (1969), in his careful study of *H. ecbasis,* estimated that up to 1300 nurse

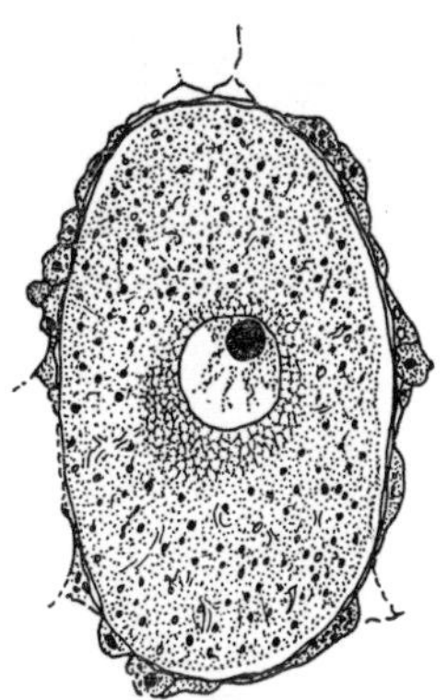

Fig. 4.1 Early oocyte of *Polymastia granulosa* (Hadromerida).
The large nucleus, prominent nucleolus and vesicular perinuclear cytoplasm are characteristic of oocytes in the Hadromerida and Axinellida.

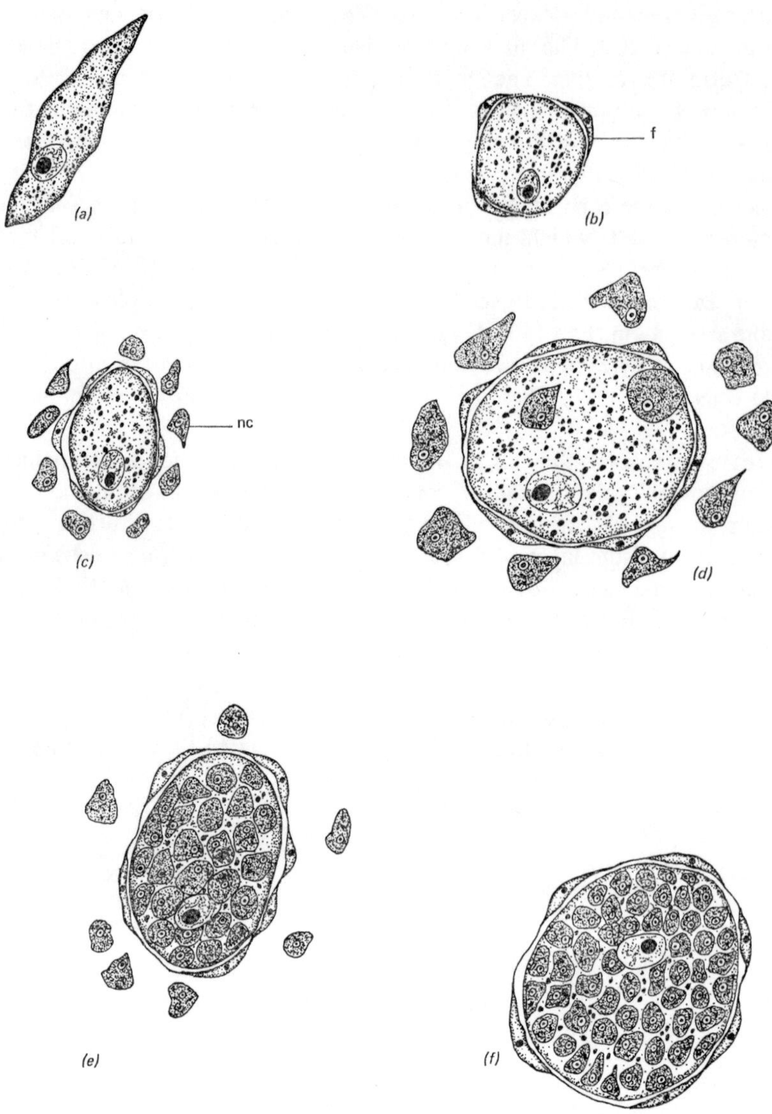

Fig. 4.2 Early stages in oogenesis in a viviparous sponge.
(*a*) Mobile oocyte; (*b*) young oocyte enclosed within a layer of follicle cells (f); (*c*) the beginning of nurse cell (nc) aggregation; (*d*) incorporation of nurse cells into the growing oocyte; (*e*) oocyte cytoplasm almost obscured by nurse cells; (*f*) nurse cell incorporation complete. (Redrawn from Fell 1969.)

cells can be accumulated by one oocyte. The sequence followed in *Hymeniacidon perleve* agrees in detail with that described for *Haliclona ecbasis*. Only the size of oocytes differs and this can be expected to vary for every species. In *Hymeniacidon perleve*, nurse cells suddenly appear throughout the sponge endosome soon after oocytes become surrounded by follicular cells. Very rapidly, streams of nurse cells with characteristic granular cytoplasm, converge on and surround the oocytes and soon form dense aggregates which obscure the oocytes. It seems clear that at a specific point in oocyte growth there is a chemotactic response by a high proportion of the archaeocyte population to a substance produced by oocytes. This promotes the differentiation and aggregation of nurse cells, which are not obvious as a special type of archaeocyte, except when the sponge is reproducing.

Meiotic division in the cycle of oocyte development is very difficult to observe. In the few cases where maturational divisions and formation of polar bodies have been observed, they occur at the end of oocyte growth, except in *Hippospongia communis* (Tuzet and Pavans de Ceccatty 1958) where meiotic divisions precede the period of major oocyte growth.

In the case of viviparous development, the oocytes must be fertilized, *in situ*, in the mesohyl. This raises the problem of sperm release and subsequent transport to the oocyte. Release of sperm has been observed several times. In the skin-diving literature, all reports of 'smoking sponges' are probably observations of sperm release. This can be synchronized in a population as, for example, in *Neofibularia nolitangere* (Poecilosclerida), or it can be restricted to individuals, as in *Verongia archeri* (Verongida) (Reiswig 1970). Clouds of sperm are emitted from the oscules over periods of up to 20 min. Where spawning is epidemic in the population, initiation of sperm release spreads down-current, suggesting that a chemical signal may trigger the release.

Once in the sea, spermatozoa can be taken into the oocyte-containing specimens in the inhalant water current. They must then cross the cellular barrier constituted by the feeding layer, enter the mesohyl, locate the oocytes and, if the species is one where a complete follicular investment encircles the oocyte, pass that last barrier, to enable fertilization to occur.

In all cases where this elusive process has been observed, it involves first trapping of sperm by choanocytes, then enclosure of sperm within a vesicle in the choanocyte. This cell then looses its collar and flagellum and migrates from the flagellated epithelium, through the mesohyl, to the ovocyte. The migratory choanocyte is termed a carrier cell (Fig. 4.8, p. 113). This carrier cell makes contact with the oocytes before follicle development is complete. The sperm, which appears as two masses (the acrosomal and intermediate segments), is transferred, in its vesicle, from one cell to the other. In nearly all cases the sperm undergoes little change in the cytoplasm of the oocyte until the maturation division occurs, then the male and female pronuclei develop and fuse. At this stage it is extremely difficult to differentiate between late oocytes and young embryos unless multiple nuclei can be distinguished.

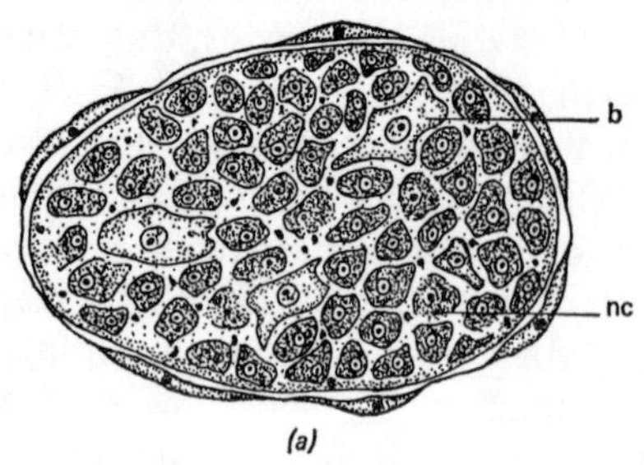

(a)

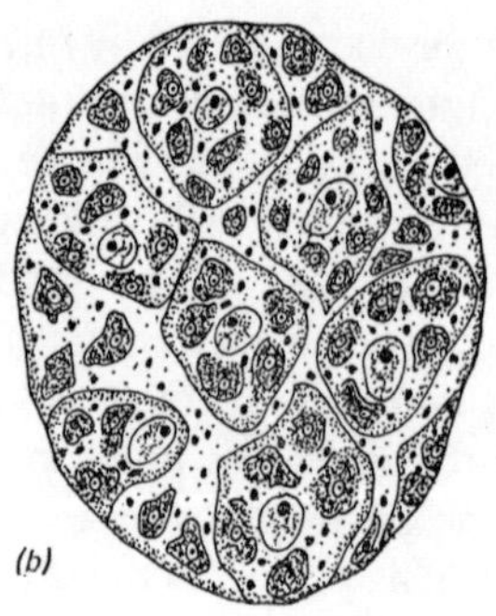

(b)

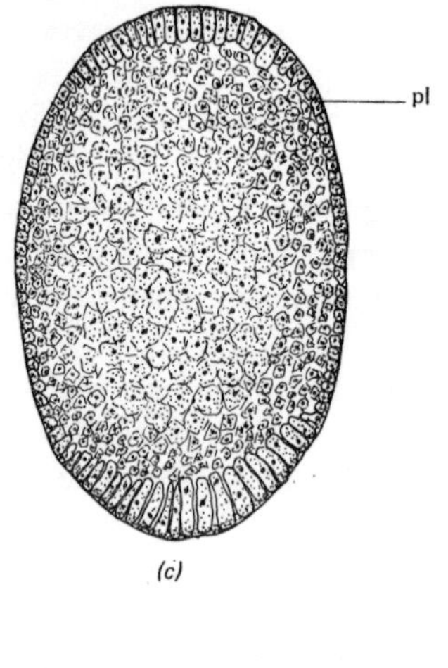

(c)

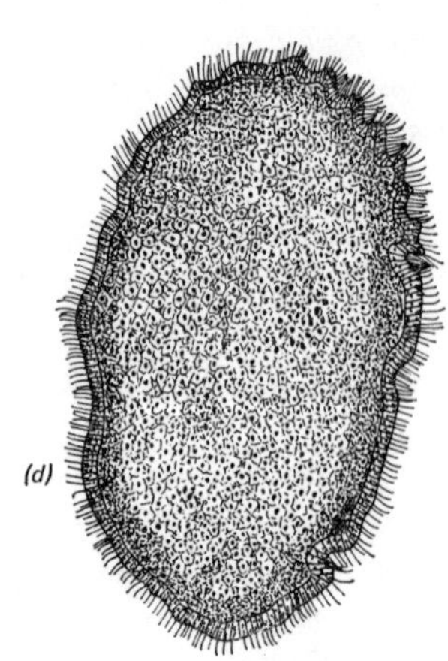

(d)

Fig. 4.3 Embryo formation in a viviparous sponge.
(*a*) The products of early cleavage are blastomeres (b) which are often almost obscured by the nurse cells (nc).
(*b*) Late cleavage, where most nurse cells are incorporated into blastomeres.
(*c*) Cellular rearrangement after cleavage is completed produces a peripheral layer of columnar cells (pl) surrounding the central mass.
(*d*) Completed parenchymella ready for release. The peripheral layer is now ciliated and the central mass contains many distinct cell types.
(*a* and *b* redrawn from Fell 1969.)

In *Haliclona ecbasis* multiple, nucleolate nuclei, often incompletely segregated by cytoplasmic partitioning, can be seen within the great mass of granular, sometimes anucleate nurse cells (Fig. 4.3*a*). Early cleavages can be regular or irregular but are always difficult to observe. The blastomeres, as they form, are filled with engulfed nurse cells. These are fragmented progressively but remain little changed; their contents are gradually incorporated into the blastomere cytoplasm. In late cleavage stages, fragmented nurse cells have disappeared and yolk granules which remain in the blastomeres are being broken down (Fig. 4.3*b*).

Differences in nurse cell incorporation during cleavage occur from species to species. Anucleate cell fragments do not retain their integrity after fusion with the oocyte in some species, but their yolk granules always pigment the blastomere cytoplasm until late cleavage and often throughout larval life.

Beyond this stage there is great variation in development depending upon the form of the larva to be produced, and the degree of cellular and skeletal differentiation to be attained before the larva can be released.

To generalize, after yolk material has been incorporated completely, cells of two types can be identified among the blastomeres: small cells with granular nuclei are dispersed among the larger cells with vesicular nucleolate nuclei. These small cells will later migrate to the periphery of the embryo where they will become ciliated and form the larval epithelium (Fig. 4.3*c*). Cell division continues during the differentiation and segregation of the different cell types and, since the small cells divide more rapidly, the superficial epithelium is very quickly organized. This epithelial layer is tightly packed, resembling a columnar ciliated epithelium. Cell differentiation to produce more of the types found in the adult proceeds in the solid, central region of the parenchymella larva (Fig. 4.3*d*).

This type of reproductive sequence, with many qualifications in detail, characterizes Demospongiae belonging to the subclass Ceractinomorpha. The accounts most frequently at variance with this interpretation are those of Tuzet (1947) dealing with *Reniera elegans* and many accounts of fresh-water sponge development. The differences are in detail, however; the general pattern of sexual reproduction is comparable.

4.2.2 Larval release and behaviour

It is not known how larvae, ready for release, leave their follicle and traverse the mesohyl to an excurrent canal. It would not be surprising to find that larvae produce enzymes which assist in this process. Another possibility is that there is a great inflation of the excurrent canals to compress the tissue and thus provide some mechanical stimulus for larval release. Certainly some sponges are inflated at the surface during the release period (Bergquist, Sinclair and Hogg 1970). Larvae can rotate and swim inside the sponge. Once expulsion begins it proceeds steadily, at a rate of four to five larvae per minute in *Microciona coccinea* and, in the same species, continues for three to four days. Larvae remaining after this time degenerate.

After release, the larvae of most sponges swim for a period which can be from 3 hours in *Microciona coccinea*, to as long as 48 hours in *Mycale macilenta*. Other larvae, although completely ciliated, have no swimming period at all, they sink to the substrate after expulsion and creep until settlement, which in *Halichondria moorei* is achieved within 20–60 hours. Parenchymella larvae can show differential pigmentation, and can have cilia of differing length in specific regions; internally they can be very simple in terms of cell differentiation (*Halisarca*) or very complex (*Mycale*). However, apart from the ciliated epithelium, they have no organization. Despite this fact they can show marked photoaxis and geotaxis and often reverse their response to light stimulus and gravity as metamorphosis approaches. There is clearly such a thing as larval behaviour in sponges, although the controls of such behaviour are difficult to discern unless they lie at the biochemical level. The fact that species-specific behaviour patterns in swimming and responses to light and gravity do occur in larvae of Demospongiae raises the possibility that sponge distribution, in what are frequently very specific habitats, may not result only from selective mortality following unselective settlement. Larval behaviour is clearly a factor in adult distribution. Preliminary studies correlating larval behaviour and adult sponge ecology support this viewpoint (Levi 1956; Bergquist and Sinclair 1968).

4.2.3 Settlement and metamorphosis

As settlement approaches the larvae enter a short creeping phase, lasting for 2–3 hours; this can be interrupted by resumption of swimming for a short time, but is always followed by a return to creeping. Some larvae spin slowly on the substrate while adhering by the anterior pole, others rest in an inclined position on the anterior pole while the cells flatten against the attachment surface (Fig. 4.4).

Once initiated, settlement and early metamorphosis are rapid. In *Haliclona* sp. (Bergquist and Sinclair 1968) the larva comes to rest vertically on the anterior pole and the cells of this region spread evenly over the substrate changing the shape of the young sponge to a cone on which the apex is occupied by the brown posterior pigmented cap, fringed by the ring of long cilia. Within 3 hours the apex collapses and the sponge become hemispherical. The ciliated ring disappears after 24 hours and the canal system, with single apical osculum, is functional soon after. The sponge at this time is 600 μm in diameter. It retains the single osculum and hemispherical shape until it reaches 2.0 mm diameter after 14 days.

These macroscopic events are accompanied by a rearrangement of larval cells and differentiation of additional cell types in those larvae, such as *Halisarca*, which have a very simple histology at the time of release. While there are instances of partial organization of choanocyte chambers and spongin fibres in larvae while still swimming, it is usual that metamorphosis and organization of the adult cell layers are delayed until attachment is achieved.

Around the attachment site, at which point the flagellated cells have already either disappeared or migrated into the solid central mass, collencytes stream

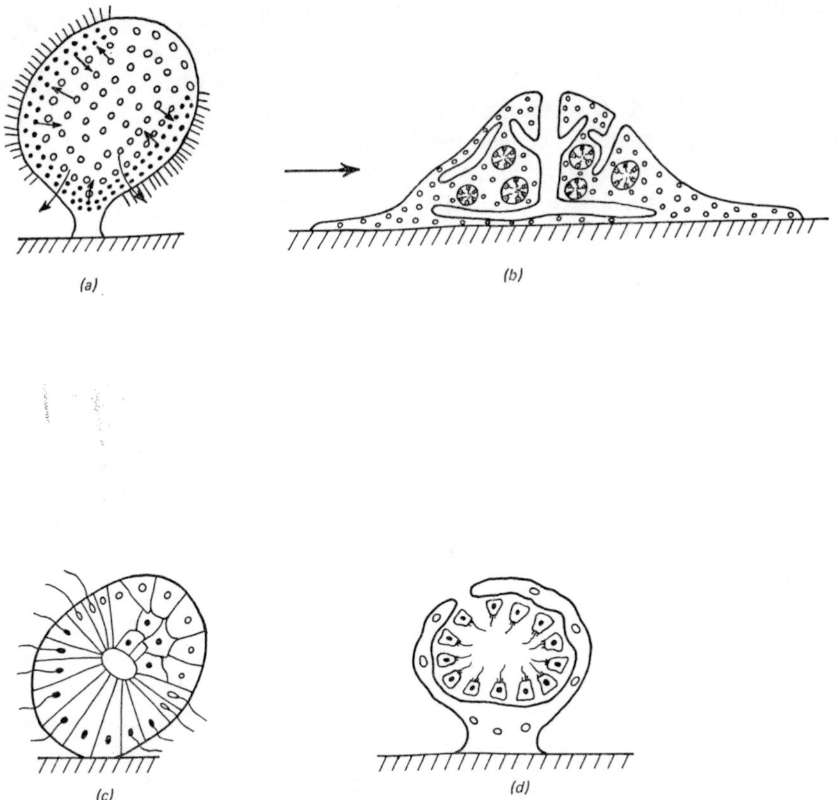

Fig. 4.4 Metamorphosis of larvae to produce young sponges.
(*a*) A parenchymella settling by the anterior pole; small arrows indicate the direction of cell movement.
(*b*) Progressive flattening produces a small sponge of the leuconoid type.
(*c*) A calcareous amphiblastula settling by the anterior pole.
(*d*) Overgrowth of the ciliated cells by the cells of the posterior pole produces an olynthus.
(Redrawn from Fell 1974.)

outward to broaden the base of attachment and form a thin marginal membrane. This brings about the flattening of the sponge (Fig. 4.4*a*, *b*). At the same time, a network of collagen fibres and associated material is secreted between the basal epithelium and the substrate.

The process of attachment and membrane formation is primary. Meanwhile the cells of the flagellated layer lose their flagella and appear to migrate into the interior of the larva, archaeocytes migrate to the periphery where they spread and differentiate to form a simple pinacoderm. In most species the great majority of the larval flagellated cells are thought to form the choanoderm of the young sponge; there is some loss by phagocytosis in those species where the number of larval epithelial cells is in marked excess of the number of choanocytes. This reversal of layers, as it has been termed, is a characteristic feature of development from the parenchymella larva, except in species where differentiation is extremely precocious such as *Ephydatia fluviatilis* (Brien and Meewis 1938). In *E. fluviatilis* choanocyte chambers are found even in unliberated larvae and represent in, that species, direct differentiations from archaeocytes, the larval flagellated cells are phagocytosed at settlement. There is need for more critical work to establish that reversal of layers does indeed occur. It is a peculiar process, the occurrence of which distinguishes sponge development as distinct from that of other Metazoa.

Beyond a requirement by a few species that the settlement surface has already been covered with a bacterial and algal film, there is no evidence for substrate selection by sponge larvae. In the case of boring sponges belonging predominantly to the family Clionidae, there is a survival requirement that settlement must be on a calcareous substrate.

Sometimes, at settlement, fusion of larvae can occur. It has been suggested that these larvae need not be from the same parent; thus fusion is a tactic which could allow the formation of heterogenomic specimens. Normally, fusion takes place between members of a clone. This behaviour has never been observed to occur in marine sponges under natural conditions. Indeed it would be almost impossible to demonstrate outside the laboratory. The fusion of recently settled or recently germinated sponges of the same species is, however, well documented for fresh-water sponges (Van der Vyver 1970). The same author has drawn attention to the opposite case, the non-fusion of larvae and metamorphosed sponges from the same species, but from different parents. It is not at all uncommon to observe individuals of the same species abutting on one another, but having quite separate pinacoderms. Van der Vyver has developed the idea of physiologically distinct strains within sponge species, with members of a particular strain having cell-surface-active factors which permit or prevent fusion and thus operate as a recognition mechanism which distinguishes self–non-self.

4.2.4 Alternative developmental strategies

The preceding account of sexual reproduction was based, with a few exceptions, on the process as it takes place in Demospongiae in which the result of reproduction is an incubated parenchymella larva. This type of development is viviparous, and at the time of release the larva can be a highly differentiated little organism. It should be noted that the use of the term viviparous is perhaps not strictly correct in the sponge context: incubation, which does not imply the transfer of maternal material to the embryo during embryonic development, more accurately describes what is known of the process. However, size increase of considerable extent can take place during development from mature oocyte to late embryo; this suggests that maternal cells are incorporated during cleavage as well as during oocyte growth. The passage of some material from parent to embryo is necessary to the strict definition of viviparity.

A majority of sponges are viviparous; however, not all viviparous forms produce parenchymella larvae. Some Demospongiae, e.g. *Plakina* and *Oscarella* (Order Homosclerophorida) and all Calcarea, produce amphiblastula larvae (Fig. 4.10*h*, p. 116). These are forms which at some time in larval development have a hollow central cavity; some are very simple, for example in the Calcinea, and have been termed coeloblastula larvae (Fig. 4.9*h*, p. 115).

Within the Demospongiae, there are some species in which larval production appears to have been deleted except for the questionable, transitory production of a patch of apical cilia on an otherwise solid cell mass. This occurs in the genus *Tetilla* in which reproduction has been observed in four species. In one species, *T. serica*, Watanabe (1957) reports that eggs are extruded inside a jelly coat, fertilized externally and then attached to the substratum. There is no swimming stage. About 48 h after fertilization the cells which occupy the site of adhesion migrate to the interior of the embryo, possibly a process homologous with the 'reversal of layers' described in parenchymella larvae. Twenty-four hours later, sheet-like projections appear on the periphery of the embryo and develop to form the future root tuft of the adult. This process has been equated with metamorphosis (Endo, Watanabe and Tamura-Hiramoto 1967). In this case

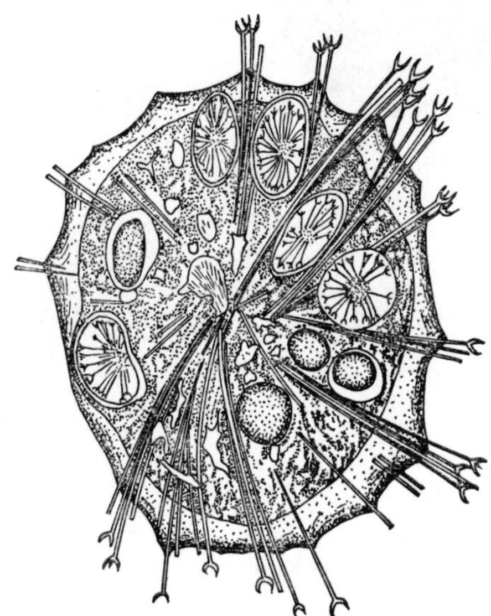

Fig. 4.5 A transverse section of *Tetilla schmidtii* to show the organized young sponges (emb) still within the choanosome of the parent sponge. (Redrawn from Sollas 1888.)

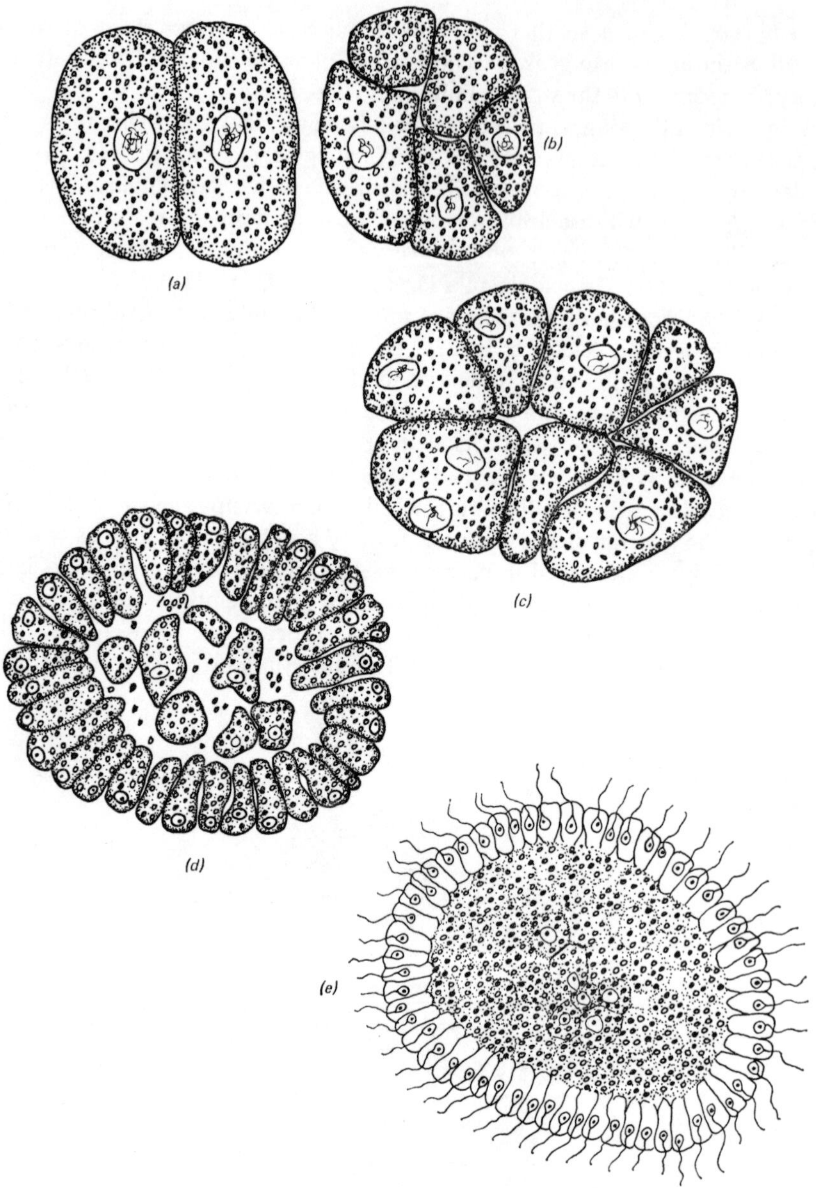

Fig. 4.6 Development of the parenchymella larva of *Tethya aurantium*.
(*a*) (*b*) (*c*) Regular and equal cleavage to produce a morula in which the blastomeres surround a small cavity.
(*d*) The cavity of the blastula is gradually obliterated by vitelline material. Some blastomeres also become internal.
(*e*) The larva is a simple parenchymella with no cavity.
(Redrawn from Levi 1956.)

development is oviparous. However, in *T. schmidtii* (Sollas 1888), *T. australe* (Bergquist 1968) and *T. cranium* (Burton 1931), development is viviparous. No larvae are produced but entire, perfect young sponges are incubated within the endosome of the parent (Fig. 4.5). This internal production of miniature sponges in which the canal system is organized and the radiate skeletal pattern of the adult already established, is also seen in some species of *Tethya* (Hadromerida) (Bergquist *et al.* 1970), while other species of the same genus are oviparous and produce parenchymella larvae (Levi 1956) (Fig. 4.6*a–d*). Regrettably, while it is possible to generalize for large groups of sponges and state that development is either viviparous or oviparous, there are some instances where both conditions are to be found within one genus.

It is difficult to speak of a typical case of oviparous development, since only four relatively complete developmental sequences have been observed. They are for *Tethya aurantium* (Hadromerida) (Fig. 4.6) and *Raspailia pumila* (Axinellida) (Levi 1956), *Polymastia robusta* (Hadromerida) (Borojevic 1967), and *P. granulosa* (Bergquist *et al.* 1970). In all cases, the time elapsed between the expulsion of the eggs and the development of the free larvae is approximately 24 hours. In *Raspailia* and *Tethya* the oocytes are not retained in any way but develop in the sea water; in both species of *Polymastia* a thick mucus is exuded with the ovocytes, and development through to the larval stage takes place in this layer. Cleavage of the fertilized egg is total and equal, leading, in the case of *Tethya*, by way of a hollow pre-swimming stage to a typical parenchymella larva. It is not clear in *Raspailia* what the nature of the free stage is, but there is no cavity between the blastomeres at any stage and the late pre-swimming stage resembles a typical parenchymella. *Polymastia* produces a flattened blastula larva, formed of equal, flagellated blastomeres. This larva crawls on its flattened surface for up to 20 days before fixation; during this time it never swims, and no cellular differentian proceeds. At settlement the larvae gradually establish a basal epithelium to ensure adhesion and these basal cells are the first to differentiate but, at the time of settlement, they still contain vitelline material. Flagellated cells lose their flagella, but do not migrate to form the adult choanocytes, they remain effectively blastomeres, still full of nutritive reserves. Choanocytes differentiate from archaeocytes at a later stage (Borojevic 1967). The benthic habit of *Polymastia* larvae recalls that of *Halichondria moorei* larvae, but the low level of cell differentiation at the time of release and the long free life make them unique among sponge larvae.

Reiswig (1976) has recorded the expulsion of oocytes in four species of Axinellida from the Caribbean. In all cases the oocytes are released synchronously in great numbers and are enrobed in heavy strands of mucus which festoon the sponge. The mucus envelope referred to earlier in *Polymastia* is not nearly as obvious as that figured by Reiswig for *Agelas* (Pl. 4*c*). Now that oocyte release in this group has been identified clearly, and since it is such a spectacular phenomenon, it should not be long before full details of axinellid development and larval structure become available.

4.2.5 Development of Hexactinellida

Only one relatively complete account is available of hexactinellid reproduction, that of Okada (1928) dealing with *Farrea sollasii.* The product of development is a parenchymella larva, incubated until a late stage within the parent sponge. The level of skeletal and cell differentiation is advanced before the larva is liberated, but at no stage have flagella been observed on the marginal cells. It is not known whether the larvae do become flagellated, or whether they are released onto the surface of the adult, and, still retaining some surface contact with the dermal membrane, move to the base of the parent sponge.

Interesting features of hexactinellid development are that sperm arise from archaeocytes, that the beginning of cell differentiation is discernible at the morula stage, that the spicule skeleton is organized very early (Fig. 4.7) and that choanocyte chambers, lined by rounded amoeboid cells, are established in the larva. It is thought that the collar and flagella of the choanocytes are not differentiated until the larva is liberated. It would be interesting to confirm and complete this account of hexactinellid development, but their deep-water habitat renders most species difficult to study. Examination of spatial distribution of adults in the shallow waters of the Antarctic reveals a high incidence of aggregation of individuals of *Rossella racovitzae* (Dayton, Robilliard and Paine 1970). Could this be a result of the deletion of larval swimming?

4.2.6 Development of Calcarea

The universal product of sexual reproduction in the Calcarea is a hollow blastula which swims for a time after release. Fertilization is always internal and the larva is incubated until maturity.

In many details of development the Calcarea are strikingly distinct from the Demospongiae on which the preceding general account was based. Fortunately, the Calcarea have been well studied, and excellent, detailed accounts exist of development in all three orders: Calcinea (Tuzet 1948; Borojevic 1969), Calcaronea (Duboscq and Tuzet 1937; Tuzet 1973*a*) and Pharetronida (Vacelet 1964).

Recent studies of calcareous sponges suggest that both the oocytes and spermatocytes derive from choanocytes, and thus ultimately from flagellated larval cells. Spermatogenesis has not been studied extensively, but it occurs in spermatic cysts dispersed in the mesohyl. Oogenesis, according to some authors, involves the movement of oogonia from beneath the choanoderm into the choanocyte chambers, where they undergo two successive divisions which result in the production of the oocytes. These then leave the chambers and take up positions under the choanocytes. This sequence has been described in representative species of all three classes. Some workers find no evidence for oogonial generations and assert that germ cells develop directly into oocytes.

Growth of the oocyte and accumulation of yolk involves the activity of nurse cells. These, in Calcarea, are either choanocytes, or cells which derive from them. They are either phagocytosed directly as in *Ascandra*, or pass nutritive materials to the growing oocytes as in *Sycon* (Duboscq and Tuzet 1937). In *A. minchini*

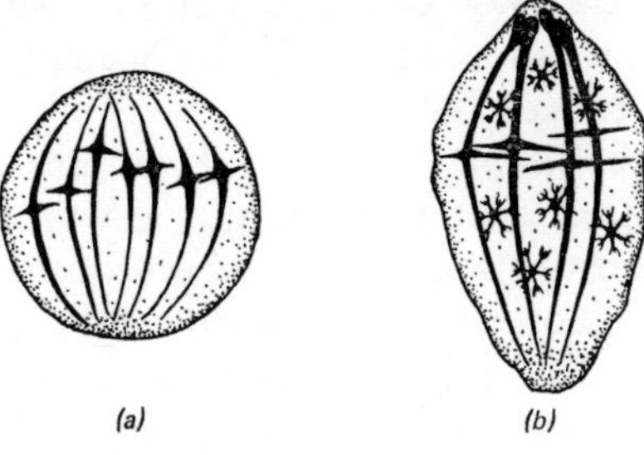

Fig. 4.7 Larva of *Farrea sollasii* (Hexactinellida).
(*a*) Megasclere spiculation developed; (*b*) microsclere spiculation added. (Redrawn from Okada 1928.)

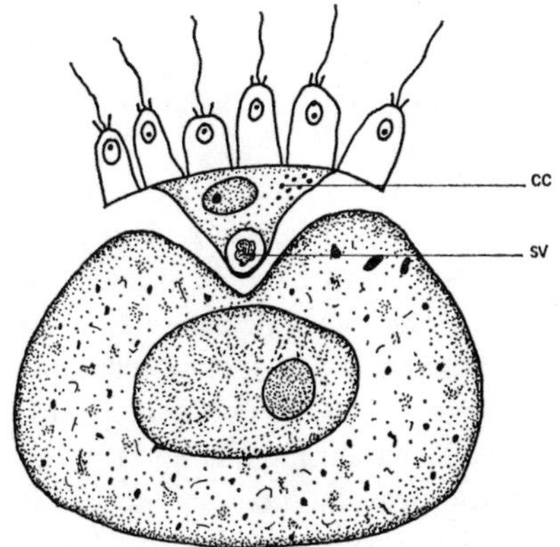

Fig. 4.8 The formation of a carrier cell in a calcereous sponge.
Sperm enter the choanocyte chamber and are trapped by a choanocyte which encloses the sperm in a vesicle (sv) and becomes migratory as a carrier cell (cc). This cell moves through the mesohyl to fuse with the oocyte. (Redrawn from Duboscq and Tuzet 1937.)

(Calcinea), after the oocyte has undergone considerable growth by the phagocytosis of choanocytes, it moves into a nest of nurse cells to complete development. As oocytes, and later embryos, develop within such a nest they are displaced through it by the influx of new cocytes and the larvae are released individually into the atrium. Nurse cells of the nests are derived from pinacocytes which grow by engulfing amoebocytes and choanocytes.

In *S. elegans* (Calcaronea) and *Petrobiona massiliana* (Pharetronida) modified choanocytes become associated with the cell that transports the sperm to the oocyte and, in this way, can supply nutritive materials to the oocyte. Maturational divisions with formation of polar bodies have been observed in a number of calcareous sponges.

Fertilization is internal in all Calcarea and the process of incorporation of sperm into a choanocyte, which then becomes a mobile carrier cell (Fig. 4.8) and transfers the sperm to the oocyte, has been observed several times. It is possible to suspend the sequence of reproductive events at the point where the carrier cell with its enclosed sperm reaches the oocyte. The former can be associated with very small oocytes, in which case fertilization can be delayed until oocyte growth is well advanced. Alternatively, fertilization takes place before, or soon after, the oocytes enter the period of major growth. Cytoplasmic continuity is established between carrier cell and oocyte and the vesicle containing the sperm is transferred from one cell to another. After this has occurred the carrier cell either separates from the oocyte or is engulfed by it. This process by which union of gametes is thought to be achieved seems to differ markedly from that which occurs in other animals and certainly it warrants further investigation. In general the sperm undergoes little change in the egg cytoplasm until the maturation divisions occur, and then the male and female pronuclei develop and fuse.

As stated earlier, the Calcarea produce more (Pharetronida, Calcaronea) or less (Calcinea) differentiated blastula larvae and in all cases the initial blastula develops directly as the blastomeres become oriented around a central space. The simplest case is that seen in some species of *Clathrina* (Tuzet 1948), where the larva is a simple flagellated bastula, a coeloblastula, with a few large, non-flagellated cells at the posterior pole (Fig. 4.9*h*). Eventually some of the cells of the flagellated epithelium lose their flagella and migrate into the central cavity where they form the choanocytes at metamorphosis (Fig. 4.9*i*). This simple process involves no reversal of layers and little cell differentiation during larval life (Fig. 4.9).

Formation of larvae in the Calcaronea and Pharetronida is more complex. In these forms, two types of cell are produced during cleavage, rounded granular cells with large nuclei, situated at the pole of the blastula nearest the parent choanoderm, and more elongate cells with smaller nuclei (Fig. 4.10*c*). The latter surround the greater part of the blastula and later develop flagella which extend into the blastular cavity (Fig. 4.10*d*). The distinctive group of rounded cells which never develop flagella are prospective pinacodermal cells, the macromeres, and for a time they do not divide, thus failing to keep pace with the flagellated cells.

Early in development an opening appears in the centre of the group of macromeres and, through this, adjacent choanocytes are ingested (Fig. 4.10*c*). This curious interlude is referred to as the stomoblastula stage. At this time the macromeres begin to multiply, the mouth closes, the smaller cells produce flagella and the embryo is a blastula with internally directed flagella (Fig. 4.10*d*).

A remarkable process of inversion now begins. This sequence is comparable

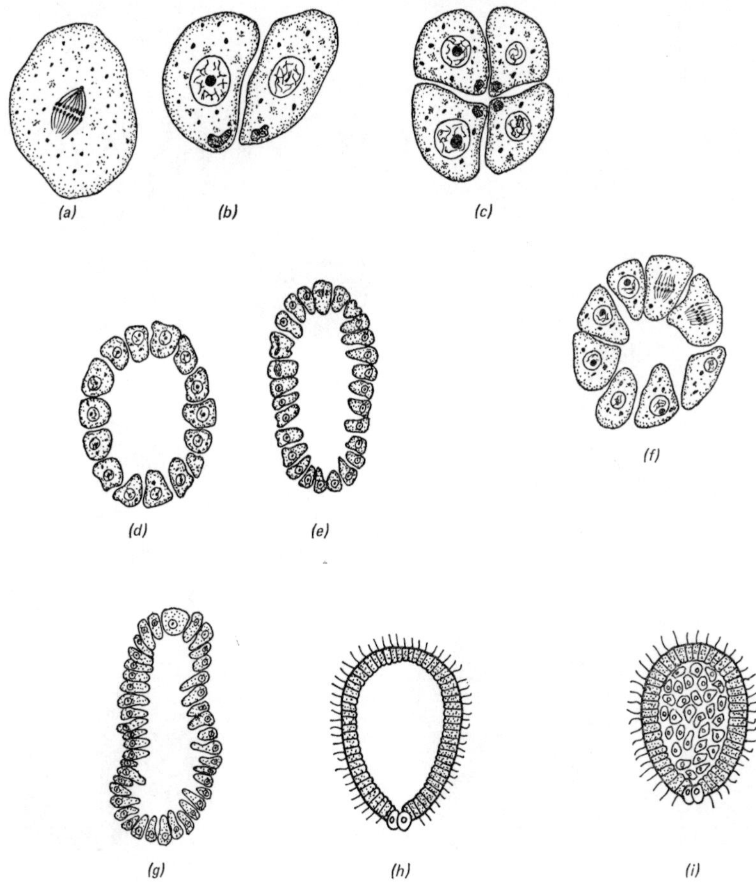

Fig. 4.9 Development of *Clathrina* (Calcarea: Calcinea)
(*a*) First division of the fertilized egg; (*b* and *c*) cleavage; (*d, e, f, g*) blastula formation; (*h*) coeloblastula with hollow interior and posterior macromeres; (*i*) late coelblastula in which some flagellated cells have moved to the interior. (Redrawn from Tuzet 1948.)

to that which takes place in the development of some green algae such as *Volvox*, but it has no counterpart either in other sponge groups, or in higher Metazoa. First, the blastular mouth reopens and enlarges to reveal the flagellated cells. These proceed to evert and the cell layer recurves until the amphiblastula, with outwardly directed flagella and posteriorally situated pinacodermal cells, is produced (Fig. 4.10*e–h*). Distributed symmetrically around the third and fourth

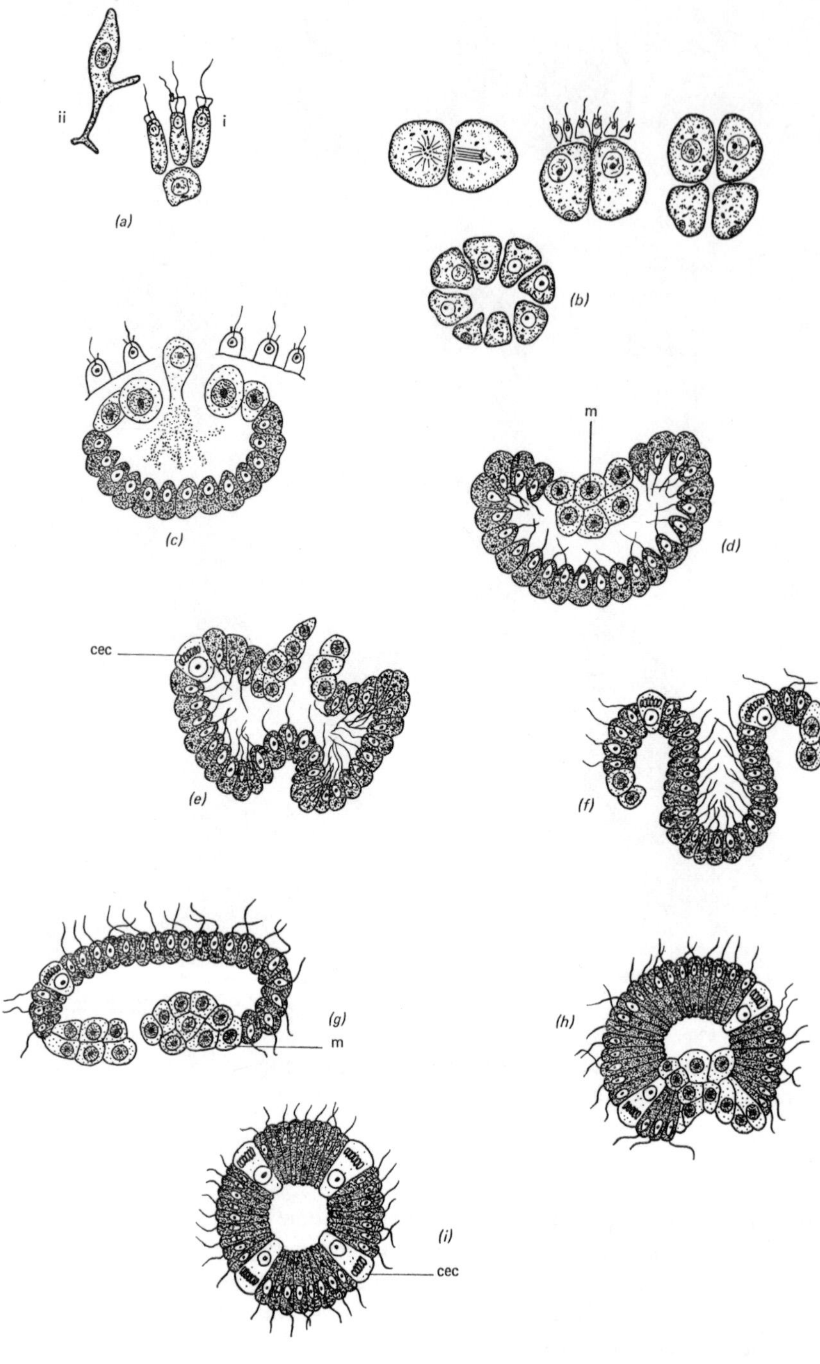

ii
i
(a)
(b)
(c)
m
(d)
cec
(e)
(f)
(g)
m
(h)
(i)
cec

Fig. 4.10 Development of *Sycon* (Calcarea: Calcaronea).
(*a*) (i) Oocyte lying in the mesohyl adjacent to the choanocyte layer.
(ii) Amoeboid, mobile oocyte.
(*b*) Cleavage stages, the embryo still lies adjacent to the choanocyte layer.
(*c*) Stomoblastula stage.
(*d*) Blastula with internally directed flagella: m, macromeres.
(*e*) Onset of inversion of the blastula: cec, '*cellules en croix*'.
(*f, g*) Two stages in the inversion of the blastula: m, macromeres.
(*h*) Longitudinal section of the amphiblastula with externally directed flagella.
(*i*) Transverse section of the amphiblastula to show all four '*cellules en croix*' (cec).
(Redrawn from Tuzet 1963.)

tier of cells above the equator of the amphiblastula are four peculiar cells which lack flagella but contain a row of chromatic corpuscles and have conspicuous nuclei (Fig. 4.10*i*). These are the *cellules en croix*. It has been suggested that they function during larval life as light receptors, but in some species the *cellules en croix* are lost before the larva is liberated. In such cases no sensory function can be ascribed to them. Regardless of function, the regular disposition of these cells in the amphiblastula is interesting as it is the only expression in sponges of any tetraradial symmetry, a feature which is so characteristic of the Cnidaria.

The embryo from the blastula stage onward is enclosed in a trophic membrane maintained by the maternal choanocytes for the purpose of supplying the embryo with food. When development is complete, the macromeres have multiplied to form approximately half the embryo (Fig. 4.10*h*). The larva forces its way through the choanocyte layer into an excurrent canal and swims to the exterior. When swimming, the larva is oriented flagellated hemisphere forward; the polar bodies are also in this area, which seems to correspond to the animal hemisphere of other metazoan embryos. On settlement, a gastrulation process ensues in which the flagellated hemisphere is invaginated, or overgrown, by the macromeres. A typical gastrula with a blastopore results in *Sycon*. The larva attaches by the blastoporal end and a small tubular asconoid sponge, the olynthus, is formed (Fig. 1.4 p. 21).

In development, the non-flagellated cells of the larva give rise to the pinacocytes and the scleroblasts, the flagellated cells become choanocytes, archaeocytes and collencytes. Mesohyl elements thus derive from both larval cell regions. This type of derivation of adult regions cannot be reconciled with conventional conceptions of germ layers and their derivatives in other Metazoa.

4.2.7 Larval types and their level of differentiation

In the preceding account of development in Porifera, I described fully developed sponge larvae of two basic types: the solid stereogastrula or parenchymella, which is similar to the cnidarian planula, and the blastula larva, usually referred to as an amphiblastula, which has a central cavity.

All Calcarea produce blastulae: these are simple types in Calcinea and more complex in Calcaronea and Pharetronida. They never show any high degree of cellular differentiation before metamorphosis, and the free larvae can still be considered to be true developmental stages.

Within the Demospongiae, parenchymella larvae predominate; they occur in all viviparous forms except those belonging to the order Homosclerophorida and often achieve a high degree of cellular diversification before release. If we ignore for a time those sponges like the Spirophorida and some species of *Tethya*, which produce no larvae, and which can be oviparous or viviparous, then the other developmental pattern to consider is oviparity, where either fertilized or unfertilized eggs are evacuated from the parent body, and development is external.

Among oviparous forms, simple parenchymella larvae may be produced, as for example in *Tethya aurantium*. These larvae, despite a relatively long free life, always remain very simple in terms of cell differentiation. They are comparable with the parenchymellae of *Halisarca* (Dendroceratida), the most primitive and undifferentiated of the incubated parenchymellae. The only other known sequences in oviparous sponges lead to the production of blastula larvae, and only in two species of *Polymastia* has the full developmental period of such larvae been observed. In this genus, throughout an extremely long free life cell differentiation is completely arrested, the larva remains the simplest of blastulae. At metamorphosis, vitelline material is still present in all cells, which are really little more than flagellated blastomeres.

It should be mentioned here that there are great problems in observing development in oviparous forms, for the fertilized eggs must be intercepted at the time of extrusion. All that is clear from present knowledge is that the end product of oviparous development is variable. It could be that further sequences remain to be described, for example, there are no data on post-gametic stages in any member of the huge order Choristida.

In the next chapter, in which sponge classification will be considered, subclass distinction will be based in some measure upon the characteristics of the larvae, and the way in which they are produced. Four quite distinct groups can be defined. In order to understand this division it is necessary to recognize the diversity of blastular types in Demospongiae and to stress the degree to which cellular differentiation proceeds in both blastulae and parenchymellae prior to metamorphosis.

The incubated amphiblastulae of *Plakina* and *Oscarella* (Homosclerophorida) are quite distinct from those of *Polymastia* and the Calcarea in that the cavity of the larva is a secondary development. The product of segmentation is a parenchymella from which the amphiblastula is derived by lysis of the internal cells. Cellular structure of these larvae remains simple, only two cell types are apparent before settlement and the larvae are still very much developmental stages. There is no parallel elsewhere in the phylum to this type of amphiblastula formation and, when other factors are considered (see Chapter 5), it is reasonable to separate the Homosclerophorida as a separate phyletic line within the Demospongiae. Another group which clearly represents a separate development is the Spirophorida with a range of unique reproductive patterns and no free larvae.

Of the remaining Demospongiae, there is a very cohesive group made up of five orders (Dictyoceratida, Dendroceratida, Haplosclerida, Poecilosclerida and Halichondrida) where incubated parenchymellae are invariably produced. These larvae, at release, contain all the principal cell types of the adult sponge, metamorphosis is merely a rearrangement of these components. Larvae of this type vary in shape, pattern of ciliation, size and degree of cell differentiation from simple forms (*Halisarca:* Dendroceratida) to extremely differentiated forms (*Mycale:* Poecilosclerida). Some fresh-water sponge larvae have organized choanocyte chambers while still free swimming. In this case the ciliated epithelium is a larval organ only and it is shed at metamorphosis, there is no reversal

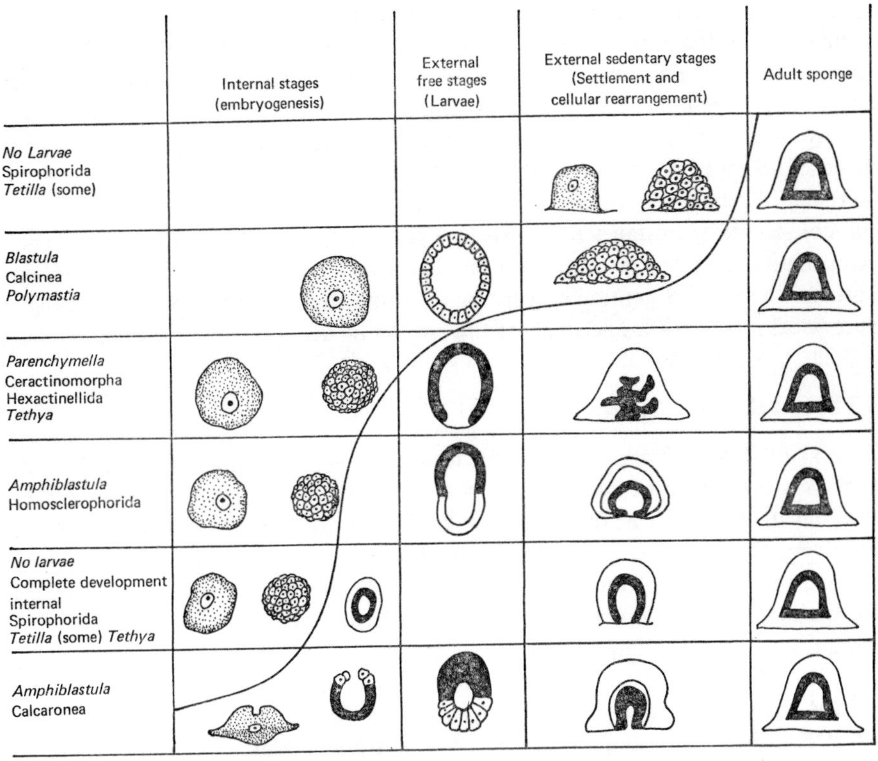

	Internal stages (embryogenesis)	External free stages (Larvae)	External sedentary stages (Settlement and cellular rearrangement)	Adult sponge
No Larvae Spirophorida *Tetilla* (some)				
Blastula Calcinea *Polymastia*				
Parenchymella Ceractinomorpha Hexactinellida *Tethya*				
Amphiblastula Homosclerophorida				
No larvae Complete development internal Spirophorida *Tetilla* (some) *Tethya*				
Amphiblastula Calcaronea				

Fig. 4.11 Diagrammatic representation of the different developmental sequences in sponges. The stages above the curve show no cell differentiation and those below are differentiated. (Expanded from Borojevic 1970*a*.)

of layers as in related forms. It is reasonable to regard these larvae as fully differentiated though unattached sponges; they are not embryonic forms. Release from the parent is delayed until differentiation is well advanced.

The orders Axinellida, Hadromerida and Choristida make up a group where development is always oviparous, but where either larval type, blastula or parenchymella, can be produced. However, during free life the larvae always remain histologically extremely simple. A theoretical arrangement of the Porifera on the basis of reproductive pattern and larval complexity is represented in Fig. 4.11.

4.2.8 Sex reversal

An interesting case of sex reversal in sponges has been recently reported by Gilbert and Simpson (1976). They determined by a programme of regular sampling in a population of *Spongilla lacustris* that the individuals were strictly either male or female in any given reproductive season. They tagged five individuals in such a way that they could be indentified reliably after their period of winter dormacy. In the following year, three of the five sponges had changed their sex, one male becoming a female and two females becoming male. Two sponges retained the sex of the previous year. This pattern constitutes a type of alternative hermaphroditism otherwise unknown in sponges, a pattern which in any one year will appear as complete sexual separation. Gametogenesis in all cases followed rapidly on gemmule hatching.

While it is only theorizing to guess at the significance of this behaviour, it is possible to see advantages for a sessile organism which disperses by free larvae. An isolated larva can produce an individual which will produce gemmules. Next spring, when those gemmules germinate, some may change sex. This enables successful fertilization and ensures the spread of the species.

4.3 Asexual reproduction

Several methods of asexual reproduction are found in sponges. Probably all species are capable of regenerating viable individuals from fragments, either residual or detached. Many species have developed more specialized methods, involving the production of gemmules of various types or the production of surface buds. Asexual reproductive devices provide mechanisms for dispersal, for maintaining attachment space and for withstanding extreme environmental conditions, in addition to their reproductive function.

4.3.1 Gemmule formation

Most references to sponge gemmules are to the complex armoured structures found in many members of the Spongillidae, the fresh-water sponges (Fig. 4.12). These small, spherical structures are produced at the onset of winter, and, invested by a thick spongin coat in which microscleres are embedded, are capable of withstanding freezing and desiccation.

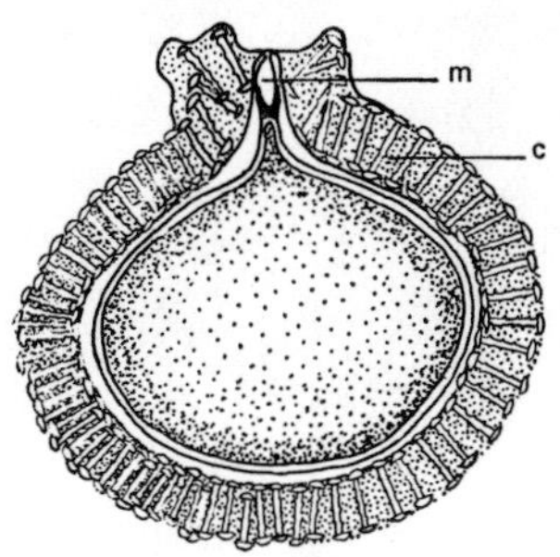

Fig. 4.12 A fresh-water sponge gemmule.
c, gemmule coat with spicules embedded; m, micropyle. (Redrawn from Evans 1901.)

The cellular events in formation of fresh-water sponge gemmules have been the subject of detailed study (Rasmont 1956, 1962). The first noticeable stage is aggregation of archaeocytes into spherical masses, 0.25–0.5 mm in diameter. Once aggregated in the mesohyl, these cells undergo active mitosis. During this period of division, nurse cells, or trophocytes as they are termed in this special context, stream toward the archaeocyte masses where they are engulfed by the active phagocytes. The result is the elaboration in the archaeocytes of vitelline platelets of very elaborate structure (Fig. 4.13). While platelet elaboration proceeds the mass becomes surrounded by an epithelium of previously amoeboid cells, spongocytes, which elaborate a three-layered spongin shell. Microscleres, secreted inside sclerocytes in the mesohyl, are transported, inside their parent cells, to the growing gemmule case and there incorporated as a spicule palisade into the spongin envelope. Both processes occur gradually, starting at one pole of the mass, and ending at the opposite pole, where, over a small circular area, the envelope remains single layered and devoid of spicules. This thinner region is the micropyle (Fig. 4.12). In the completed gemmule the envelope surrounds a dense mass of cells, so full of vitelline platelets and so tightly packed that they become polyhedric. These modified archaeocytes, usually referred to as 'thesocytes', become binucleate and, at this point, the hibernation of the gemmule commences.

When environmental conditions are again favourable for sponge growth, the gemmule hatches. This process begins at the micropylar end. Here the thesocytes divide to produce smaller archaeocytes that begin to digest their vitelline platelets. The micropyle is opened, probably by means of enzymes digesting its membrane, and the first archaeocytes flow out. These cells, which have almost depleted their reserves, spread over the gemmule case onto the substrate where they build the pinacocyte epithelia and collencyte network of the sponge. The

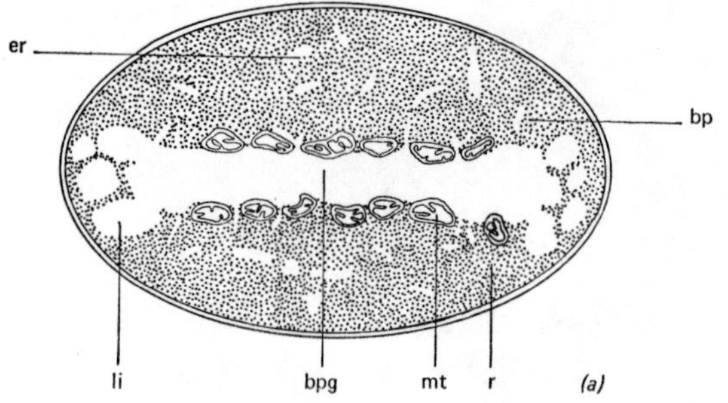
er
bp
li
bpg
mt
r
(a)

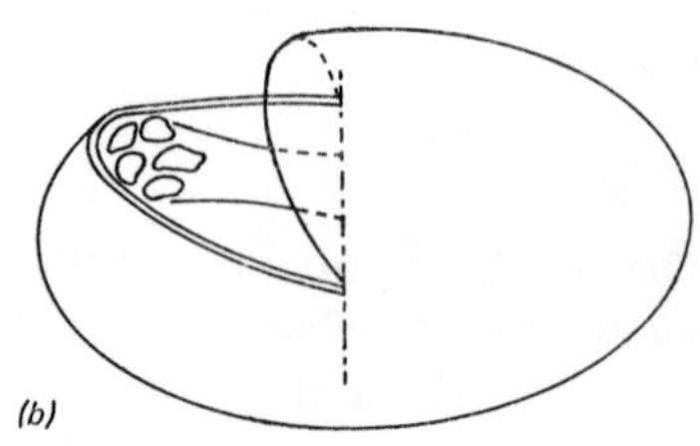
(b)

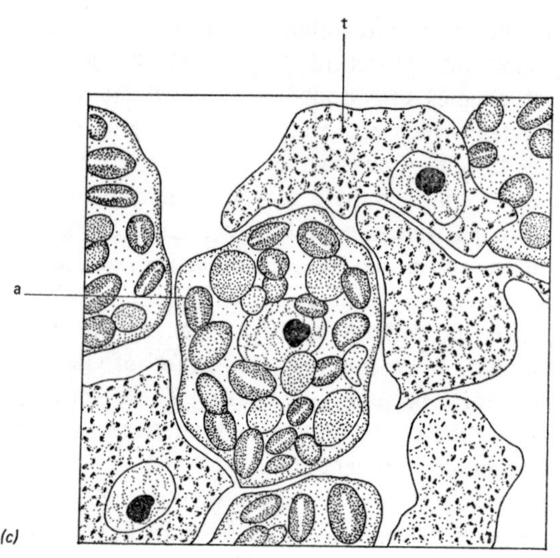
t
a
(c)

Fig. 4.13 Vitelline reserves of fresh-water sponges.
(*a*) Diagrammatic section of a vitelline platelet as revealed by electron microscopy and histochemical tests.
bp, basic proteins; mt, mitochondria; er, granular endoplasmic reticulum; li, unsaturated lipids; bpg, basic proteins and glycogen; r, ribosomes.
(*b*) Perspective diagram of a vitelline platelet.
(*c*) View of the internal cells of the sponge gemmule as seen in light microscopy.
a, archaeocyte with vitelline platelets; t, trophocyte.
(*a* and *b*, redrawn from De Vos 1971; *c*, redrawn from Leveaux 1941.)

next archaeocytes to move out of the gemmule colonize this framework, differentiate into many cell types, and produce choanocyte chambers. Finally, some archaeocytes move into the young sponge without differentiation, thus providing a reservoir of totipotent cells to meet future requirements.

This sequential difference in the fate of archaeocytes after they leave the gemmule does not appear to reflect absolute distinctions within the morphologically homogeneous cell population. Rather, archaeocytes differentiate in response to the architecture of the surface they move onto, and not as dictated by any predetermined potency. In the course of gemmule hatching, archaeocytes give rise to every cell type of the adult sponge, an excellent demonstration of the central importance of these cells in sponge morphogenesis.

No other sponges produce gemmules which are as complex as those of fresh-water sponges. However, many marine species produce asexual reproductive bodies constituted of archaeocytes which are packed with reserve material, and enclosed within spongin, or spongin plus spicule, envelopes. Such structures are found mainly in Haplosclerida (*Haliclona* sp.) and Hadromerida (e.g. *Suberites*, *Cliona*), and are the functional equivalent of fresh-water sponge gemmules. In all cases these gemmules are overwintering devices, and remain attached to the substrate after the sponge has disintegrated.

The process of gemmule formation in marine sponges differs in detail from species to species, but the general sequence is comparable to that observed in fresh-water species. Aggregations of archaeocytes form in the mesohyl, usually towards the base of the sponge, other cell types are displaced, spongocytes from the mesohyl form a spongin sheath on the outer layer of gemmule cells, or the outer layer of archaeocytes assume this function and secrete the capsule on the acellular bounding membrane. Spicules, when added to the capsule, are not inserted in precise, regular fashion (Fig. 4.14). In the resting gemmules, the archaeocytes are comparable to thesocytes, being full of lysosomes, glycogen, vitelline platelets and lipid. The micropyle is marked by a cushion-like thickening of the spongin coat.

There has been little histological study of the germination of marine gemmules although Connes (1975), working on *Suberites massa*, has established that after 6–7 days in culture the thesocytes are supplemented by smaller, non-vitelline

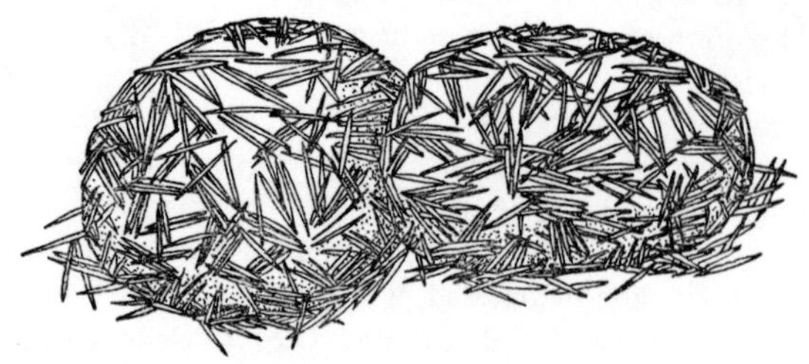

Fig. 4.14 Gemmules of a marine sponge, *Haliclona loosanoffi*. (Redrawn from Hartman 1958*a*.)

archaeocytes. After 8 or 9 days the region near the micropyle is packed with these cells which, on breaking of the coat, flow out of the micropyle, flatten quickly on the gemmule coat and on the substrate and then attach. Differentiation of the exopinacoderm is rapid, and at about 11 days, canals and spicules are established and choanocyte chambers are forming. By the fifteenth day an oscular tube has developed and choanocyte chambers are functional.

Small discrete capsules such as sponge gemmules are amenable material for experimental study and many workers have investigated the controls of gemmule production and germination. Rasmont (1962, 1970) has reported on the physiological controls of gemmulation in fresh-water sponges. His laboratory has concentrated on this research topic for many years and interesting results have been obtained. Initially the situation regarding gemmule germination appeared simple. If gemmules were gathered in late autumn and stored at refrigerator temperatures (3 °C) they could be induced to germinate by placing them in well-aerated water at room temperature. Winter dormacy of gemmules appeared to be a result of direct inhibition of development by winter frosts. Brøndsted (1936), however, had noted that germination rates of *Spongilla lacustris* gemmules could be enhanced by previous cold storage. This observation led Rasmont (1962) to compare the hatching rate of gemmules of four Belgian fresh-water sponges following different thermal treatments.

An initial comparison of two species, *Ephydatia fluviatilis* and *E. mulleri*, served to demonstrate that two quite different hibernation mechanisms existed. Gemmules of both species were gathered in early summer and placed directly into oxygenated water at 20 °C. After 10 days between 90 and 100 % of the *E. fluviatilis* gemmules had hatched and only 0–10 % of the *E. mulleri* gemmules. If gemmules of both species were stored at 3 °C for the interval between collecting and incubation, the results were quite different. In *E. mulleri* the time until the onset of germination was shortened, and the maximum rate of germination at 20 °C increased as duration of the cold pre-treatment was extended. If this vernalization period was one month, the hatching efficiency reached 90–100 %. Vernalization thus enhanced germination dramatically in this species. However, in *E. fluviatilis* a 6-week vernalization at 3 °C delayed the onset of germination, regardless of whether incubation was carried out at 12, 16 or 20 °C. At 12 and 16 °C the maximum rate of germination was also depressed. In this species it appeared that some process preparatory to hatching was slowed down by cold.

In the case of *E. mulleri*, the enhancement of this unknown maturation process by previous cold treatment was not an all or nothing process. Cold did not trigger maturation, maturation was merely accelerated by vernalization. This was demonstrated well by the fact that the germination rate of gemmules rose steadily, even when they were stored at 18 °C. After two or three months at room temperature half of the stock were ready to hatch without previous vernalization.

In summary, while *E. fluviatilis* gemmules were able to germinate immediately after formation, those of *E. mulleri* had to undergo a maturation, one stage of which could be traversed more rapidly at low temperatures. Such a stage corresponds to diapause as it occurs in many groups of organisms. In sponge gemmules, as in insects, diapause is a stage of lessened respiratory activity and lessened sensitivity to cyanide inhibition of metabolism.

In a third species, *Spongilla fragilis*, gemmules were also shown to undergo a diapause, but the optimum temperature of maturation was higher than for *E. mulleri*. The diapause was less deep. The last species studied, *S. lacrustris*, had a more complex cycle. It produced gemmules of two colours, green and brown. The green colour results from the presence of symbiotic zoochlorellae. When not exposed to previous chilling, the green stock hatched sooner, and at a higher maximum rate than the brown. After 2–4 weeks' vernalization the germination efficiency of both stocks reached about 90 %. Thus the increase in germinative ability which can be attributed to vernalization is much greater in brown gemmules than in green ones. It was at the time tempting to assume that the difference between the two gemmule types of *S. lacustris* related, directly or indirectly, to the presence or absence of zoochlorellae, but this has not been confirmed by experiment.

Ephydatia fluviatilis and *S. lacustris* produce an inhibitor of gemmule germination, gemmulostasin, but *E. mulleri* does not (Rasmont 1965; Rosenfeld 1970). The first two species produce gemmules late in the year and do not degenerate until the onset of winter conditions. The inhibitor produced by the parent tissue inhibits the germination of the gemmules until this time. The gemmules of *E. fluviatilis* and the green gemmules of *S. lacustris*, unlike those of *E. mulleri*, do not undergo diapause. The inhibitors are not species specific. They will act to inhibit germination of *E. mulleri* gemmules even after their diapause is over. In the case of *E. fluviatilis* the inhibitor acts on an early phase of germination and, once this phase is past, the inhibitor has no further effect. If the gemmule capsule is pierced, germination is not inhibited by gemmulostasin, although it is retarded compared with untreated controls (Rosenfeld 1971).

The actual formation of gemmules by fresh-water sponges is so clearly an adaptation to seasonal climatic changes, that at first it was reasonable to believe that seasonal alternation itself induced the gemmulation. However, conflicting reports on the relationship of gemmule production to seasonal temperature changes required that an alternative hypothesis be developed. Leveaux (1939) suggested that gemmulation was a byproduct of sponge growth, implying that a certain minimum size had to be attained before gemmule production. This suggestion has proved inadequate as an explanation, since, in nature, small sponges can be found gemmulating early in the year. Also, small sponges hatched in culture from a few gemmules can start building their own gemmules after a few weeks' growth, during which time temperature, oxygen concentration and illumination have all been constant. Rasmont evolved the hypothesis that gemmulation depended upon internal factors which related to sponge growth, and external factors, such as the availability of suitable food. To test this assumption, culture techniques involving standard medium and controlled diet were developed, and in such cultures whole populations of sponges of known age and initial size can now be grown under standard conditions.

From experiments in culture, Rasmont (1962, 1963, 1970) has been able to establish several facts. There is a size below which gemmulation never occurs, and a size over which it will always occur regardless of nutritive conditions. Between these two threshold values, which in numerical terms may vary from species to species, gemmulation speed increases with sponge size, irrespective of age and nutritional condition.

Gemmulation speed depends in a complex way on size, and on the composition and quantity of the sponge diet. Precocious gemmulation, or the hatching of a gemmule inside an already differentiated sponge, always increases gemmulation speed. However, this *in vivo* hatching can only occur under certain conditions; normally the parent sponge tissue exerts an inhibitory effect on any gemmule it contains, and the strength of the inhibition relates to the physiological state of the sponge.

Rasmont (1963) suggested that the observed facts could be explained tentatively if it was assumed that the onset of gemmulation was related to the accumulation within the sponge of some type of blastogenic material. Then, the triggering of actual gemmule formation would be the function of an inducer which could diffuse within a sponge, and also from one sponge to another. Inhibition of hatching would be the function of another diffusable agent, the gemmulostasin previously mentioned. A good deal of work is now being directed toward observing changes in the relative abundance of different cell types in fresh-water sponges in the period prior to gemmulation. A shift toward nurse cell and archaeocyte production should be recognizable, even before accumulation of these cells begins.

There has been some experimental work recently on the production and hatching of gemmules in marine sponges (Fell 1974) and in many respects the processes are similar to those observed in fresh-water species. The sponge studied was *Haliclona loosanoffi*, a species known to produce gemmules over a wide range of environmental conditions (Hartman 1958*a*; Wells, Wells and

Gray 1964). Gemmules are present through most of the year along the New England coast; they are produced in June or July and do not germinate until the following May or June. During late summer or early autumn the parent sponges degenerate, leaving the gemmules attached to the substrate. The gemmules are the only form in which *H. loosanoffi* exists during the colder parts of the year in the northern part of its range. They are an obligatory part of the life history.

Fell has shown that gemmules of *H. loosanoffi* may be held at 5 °C for up to ten months without a sign of germination. Such gemmules germinate readily when they are subsequently cultured at 20 °C. Germination will also occur at 15 but not at 10 °C. Well-organized sponges can develop in 1 or 2 weeks at 20 °C and 2 to 4 weeks at 15 °C. These laboratory results agree with field observations which suggest that germination normally occurs as the water temperature rises from about 15 to 25 °C. It appears that germination of gemmules is prevented during the winter by low temperature. This pattern accords with most observations on fresh-water sponges. However, like fresh-water sponges, the regulation of gemmulation is not simple. Gemmules begin to form in early summer when the water temperature is close to 20 °C, a temperature close to that in the spring when germination occurs. It is clear that some mechanism operates to prevent gemmules germinating prior to the onset of winter. Experiments indicate that, in *H. loosanoffi*, cold pre-treatment enhances germination, some process in gemmule germination requires a relatively long period of time, and this process is accelerated by low temperatures. Marine sponge gemmules also have a typical diapause.

When and how diapause is initiated in *H. loosanoffi* is unknown, whether at the time of gemmule formation or later. Also, nothing is known of the nature of the diapause itself. One observation on the germination of gemmules which were not subjected to any cold pre-treatment is interesting in this connection. Often, many gemmules situated along the cut edges of cultures germinated within a few days, while those away from the lesion did not germinate, even if left for a long time. This suggests that injury to the gemmule capsule, or to the contained cells, may stimulate germination. It is possible that the gemmule capsule, like the coats of certain seeds, is important in maintaining dormancy, either by restricting respiration, or by retarding the loss of germination inhibitors.

Haliclona loosanoffi has a gemmulation cycle similar to that of *Ephydatia mulleri*; the time of gemmulation in relation to sexual reproduction and seasonal characteristics is the same, a diapause is present, and there is no evidence in either species of production of germination inhibitors.

4.3.2 Bud formation

Marine sponges produce external buds of three types. Most common are the stalked external buds which are produced mainly in Hexactinellida (Fig. 1.5, p. 24) and Hadromerida (Fig. 4.15). Several Hadromerida, in addition to stalked surface buds, produce large buds from basal attachment stolons. In some *Tethya* sp., and *Aaptos aaptos*, buds of this type remain attached to the

Plate 5

(a) A view of the surface of a living poecilosclerid sponge showing the complex system of ostia (o) opening into underlying inhalant canals and the large oscules which just below the surface receive several exhalant canals (ec). The elevation of the oscular membrane above the sponge surface is evident. (Photo W. Doak.)

(b) A photomicrograph of the superficial dermal spongin fibre skeleton typical of members of the Callyspongiidae. Primary and secondary fibre networks are developed.

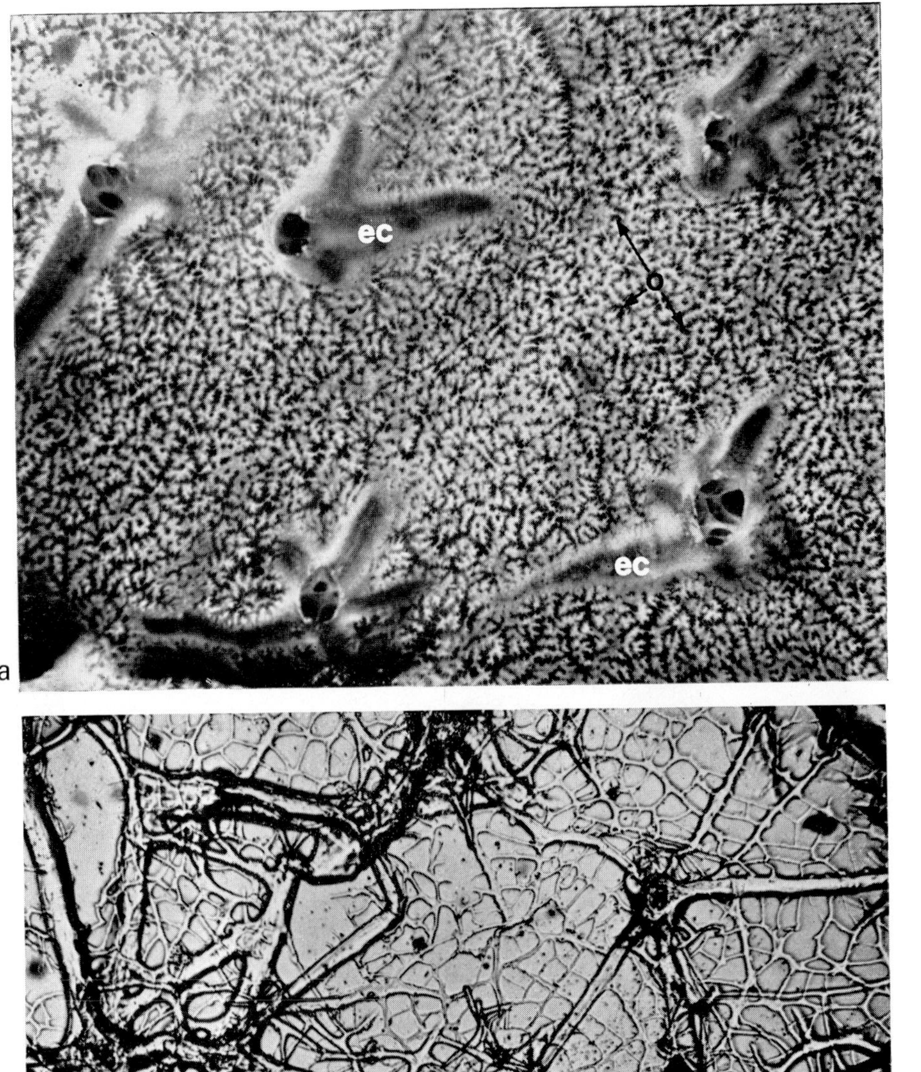
ec
o
ec
a
b

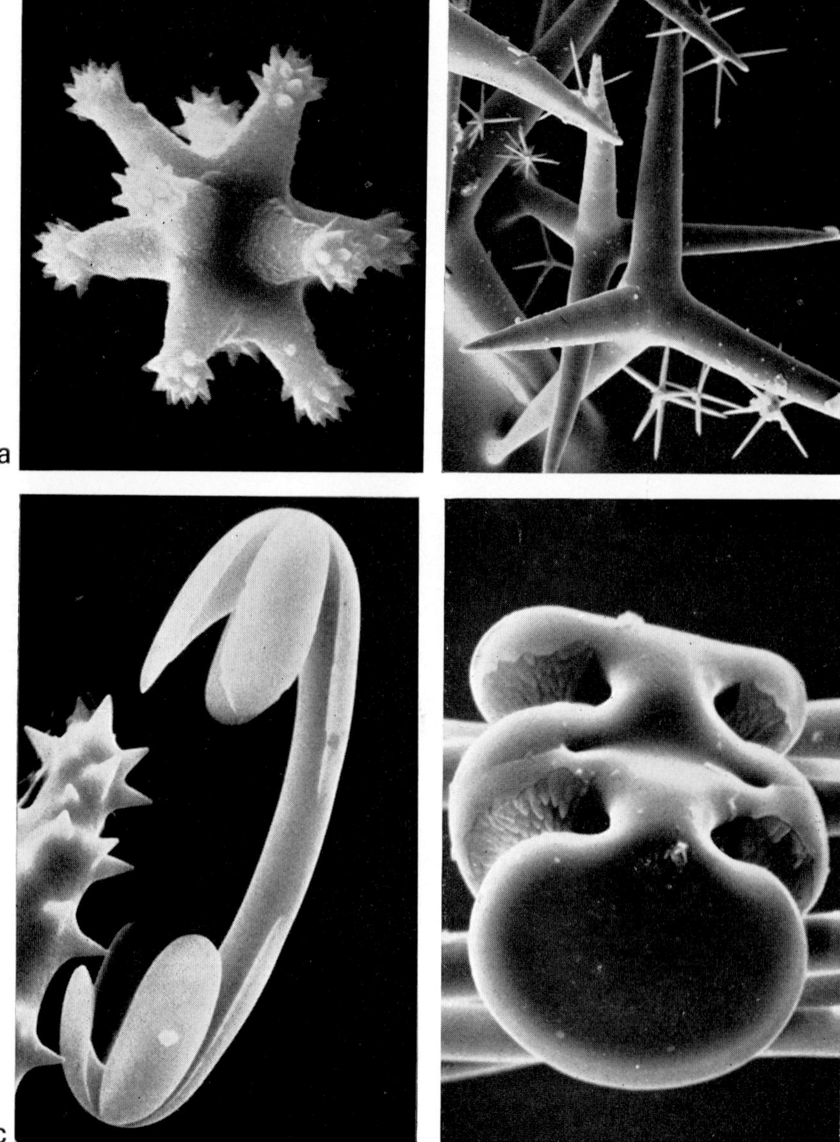
a
b
c
d

Plate 6
Scanning electron micrographs of some representative sponge spicules.

(a) An asterose microsclere from a species of *Tethya*. This spicule has a pronounced core or centrum and many short rays each with a terminal cap of spines.

(b) Tetraxon megascleres (calthrops) and asterose microscleres of an undescribed genus of the Choristida (Calthropellidae). The calthrops megasclere is typical of this order.

(c) A typical arcuate isochela as seen in *Ectoyomyxilla*. Portion of a spiny (acanthose) m gasclere, an acanthostyle is also shown.

(d) A bizarre type of cheloid microsclere as seen in *Tetrapocillon*.

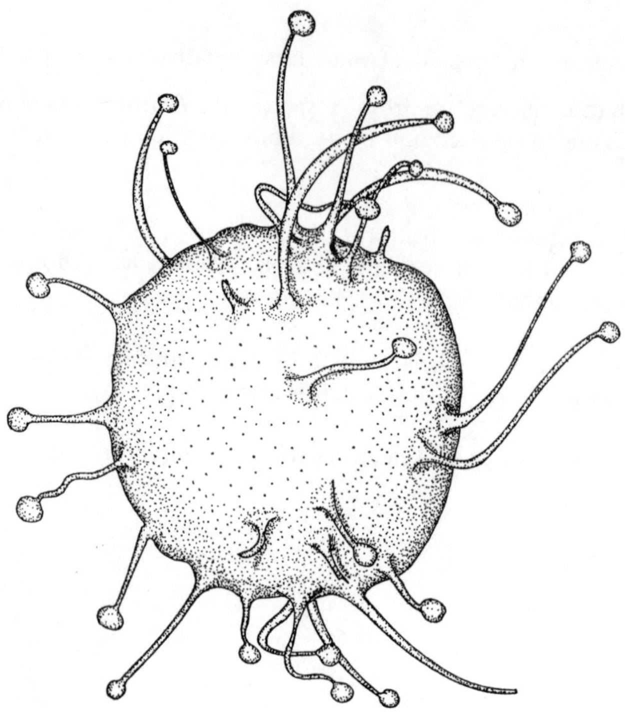

Fig. 4.15 Tethya ingalli (Hadromerida) with stalked surface buds.

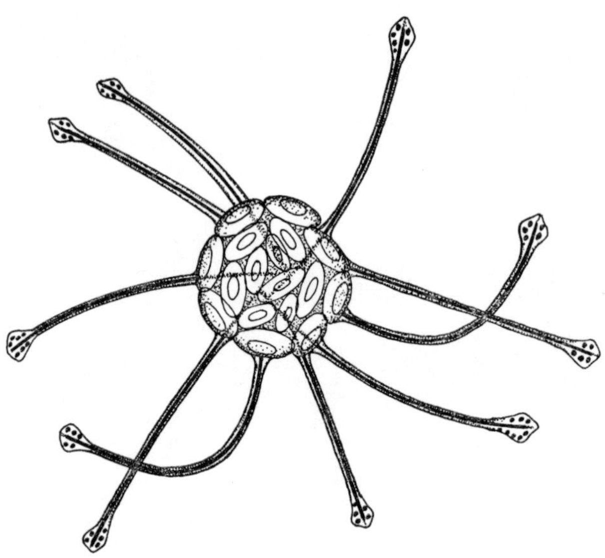

Fig. 4.16 External bud of *Alectona*. (Redrawn from Garrone 1974.)

parent sponge while the surface buds detach and become free for a short time. Finally there are the strange 'armoured gemmules' produced by two genera of the Clionidae, *Alectona* and *Thoosa* (Fig. 4.16). These structures have on rare occasions, since 1892, been collected in plankton samples, and a paper by Garrone (1974) described the ultrastructure of a single specimen and demonstrated definitively that these planktonic stages are not gemmular or larval. They have a complex histology and skeletal arrangement and an active metabolism directed toward collagen synthesis, and compare very closely in cellular composition with external buds of *Tethya* (Fig. 4.15). Apart from their remarkable morphology and spicule armour, these free buds are notable within the phylum as structures which are perfectly adapted for an extended planktonic life. Clearly, with their extensive vitelline reserves they can live for some time in the plankton, and, on settlement, are equipped for rapid reconstitution of an attached individual.

Bud formation has been best studied in species of *Tethya*. Initially buds arise as club-shaped protrusions from the sponge surface and consist of bundles of spicules surrounded by masses of archaeocytes. Mature buds are spherical knobs supported on long filaments. After the bud has been released the filaments are most frequently resorbed by the parent sponge. At the time of release the cortex and choanosomal regions have become differentiated from the bud.

In the western Mediterranean, *Tethya aurantium* is known to produce external buds all year round but with greater frequency in colder months. There are no detailed observations available from other parts of the world but *T. ingalli* in northern New Zealand has been found budding all year round.

It is very probable that many Demospongiae produce external buds which are not elevated on the long stalks which render these structures so obvious in *Tethya*. Devos (1965) has reported on the production of small spherical buds in *Mycale contarenii*. These are situated in small depressions in the surface of the parent sponge and are formed simply by the pinching off of a small portion of sponge which contains all necessary cell types. Subsequent reorganization results in the formation of a small sponge. *Stelletta arenaria* from New Zealand produces similar buds.

4.3.3 Asexual production of larvae

Since the early observations of Wilson (1891, 1894), there have been a number of reports which suggest that in some Demospongiae and Hexactinellida parenchymella larvae can be produced by an asexual process. Naturally, such a suggestion has aroused opposition: Maas (1893) immediately questioned Wilson's interpretation of the development of *Mycale fibrexilis*. The most recent paper which described asexual production of larvae (Bergquist *et al.* 1970) indicated that this phenomenon was common in intertidal Demospongiae, where it was interpreted as a supplementary reproductive process which enabled retention of a free dispersal stage even if fertilization failed.

As in many other areas of sponge biology, what was lacking in efforts to resolve this debate was a well-documented and illustrated account of development in a sponge which incubated parenchymella larvae. Bergquist *et al.* (1970) set out to provide this information, as did Simpson (1968). In both cases the data, even after up to four years of sampling, were incomplete and inconclusive and therefore difficult to interpret.

The account of development in *Haliclona ecbasis* (Fell 1969) provided the necessary basis for interpreting stages in oogenesis, segmentation, and larval production. Since that time, two New Zealand species, *Hymeniacidon perleve* and *Halichondria moorei*, have been further studied by light and electron microscopy. Regular sampling has shown that in *H. moorei* the sexes are separate. This accounts for the previously noted absence of sperm, in specimens which were obviously reproducing. This puzzled earlier observers and led to their attributing development of larvae to asexual processes. In addition, in the extremely dense 'gemmular stages' where no blastomere nuclei had been discerned in the early study (Bergquist *et al.* 1970), electron microscopy has now shown that they are present. Thus all 'gemmular' stages, thought to have arisen from archaeocyte aggregations, are now, in this species, interpreted as heavily vitelline equivalents of the *Haliclona ecbasis* embryo at about the 1600 cell stage, or later.

Careful search for ovocytes in *Halichondria moorei* has revealed that they are present from January to July, always in low numbers. There is no need therefore to invoke asexual processes in larval production in this species. It is probable that detailed study will reveal other intertidal species to be similar. An added note is that in species which are reproducing all the year round, there are periods in which embryo development slows or stops, only to resume later in the season. This leads to the accumulation in the sponge of many different developmental stages at any particular time, and makes developmental sequences difficult to interpret.

4.4 Reproductive cycles

To establish the reproductive cycle for particular sponges requires regular sampling of a population over a long period of time. Consequently there are relatively few species for which gametogenic cycles are known.

In the oviparous sponges from the Mediterranean, *Stelletta grubii, Axinella damicornis, A. verrucosa,* and *Polymastia mammillaris,* there is a definite reproductive period with one cycle of oocyte production per year. Oocytes appear several months before spermatocytes and only when ovocytes are fully developed do spermatocytes begin to differentiate (Fig. 4.17).

In the case of *S. grubii*, individuals are asynchronous with respect to sperm differentiation: in *P. mammillaris* and the two species of *Axinella*, sperm differentiation is synchronous within individuals and throughout the population. Sexes are separate in *S. grubii* and in both species of *Axinella*; in *Polymastia* oogenesis precedes spermatogenesis in the same individual.

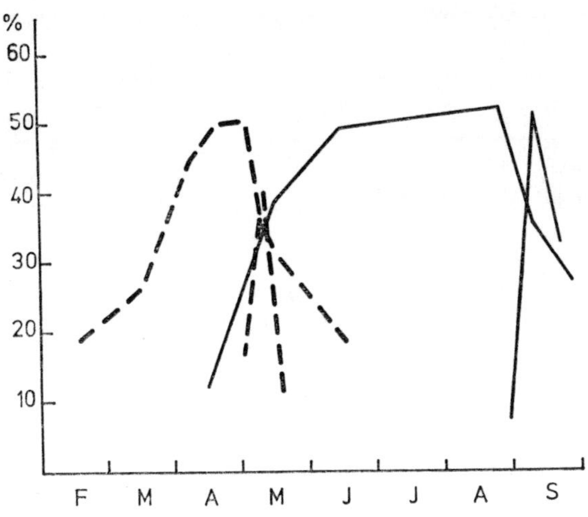

Fig. 4.17 Sexual reproductive periods of two species of *Axinella* which share the same habitat in the Mediterranean. Periods of gamete production are well separated. The percentage of the population engaging in ovogenesis or spermatogenesis is plotted against the month of the year. Dashed lines, *A. damicornis*; solid lines, *A. verrucosa*. Steep peaks, spermatogenesis; extended curves ovogenesis. (Redrawn from Siribelli 1962.)

In the calcareous sponges, *Petrobiona massiliana, Sycon* and *Leucandra*, oogenesis is synchronized within populations. Most specimens of a species examined at a particular time and location contain oocytes or embryos at the same stage of development. In the same calcareous sponges, there may be a succession of reproductive cycles leading to the coexistence of oocytes and embryos in the same specimen. When this happens, there is sometimes topographical separation of the older and younger embryos within the sponge.

A common pattern is that shown by *Haliclona ecbasis* (Fell 1970), where the sponge has a definite reproductive period with repeated, asynchronous cycles of oocyte and spermatozoan differentiation (Table 4.1). Frequently, particularly in warm temperate regions, individuals or local populations of a species may reproduce throughout the whole year, but in such cases one or two periods of maximal larval production are obvious (Bergquist and Sinclair 1973).

In the cases of *Haliclona loosanoffi, Halichondria bowerbanki* and *Microciona prolifera* the inter-relationship of sexual reproduction and hibernation on a seasonal basis has now been worked out (Table 4.2). Preliminary work by Fell (1974) indicates that, in *Haliclona loosanoffi*, only adults grown from gemmules are able to reproduce the following summer. The products of growth following larval settlement presumably produce gemmules and then disintegrate. They will be the source of the following year's sexual production. *Haliclona loosanoffi* in the southern part of its range (North Carolina) has a strikingly different

Table 4.1 *The occurrence of reproductive elements in* Haliclona ecbasis *in San Francisco Bay.*

Year*	Month	Sperm	Small oocytes	Large oocytes	Embryos	Larvae
1964, 1966	January	?	+	+	+	?
1964	March	?	+	+	+	?
1964	April	?	+	+	+	?
1964, 1966	May	?	+	+	+	+
1965	June	+	+	+	?	–
1965	July	+	+	+	+	+
1965	August	+	+	+	+	+
1964, 1965	September	+	+	+	+	+
1964, 1965	October	+	+	+	+	+
1964, 1965	November	+	+	+	+	+
1965	December	–	+	+	+	+

*The sponge was absent (or extremely scarce) at the collecting stations during the summer of 1964 and the first several months of 1965, and few collections were made in 1966. (After Fell 1974.)

reproductive cycle. In Hatteras Harbor, North Carolina there are large annual changes in water temperatures, but the levels are substantially warmer than along the New England coast. *Haliclona loosanoffi* has two reproductive periods, in June and July and then in October and November. The sponge lives over the winter in a non-reproductive state and survives the warm summer months (late July to early September) only in the form of gemmules (Fig. 4.18).

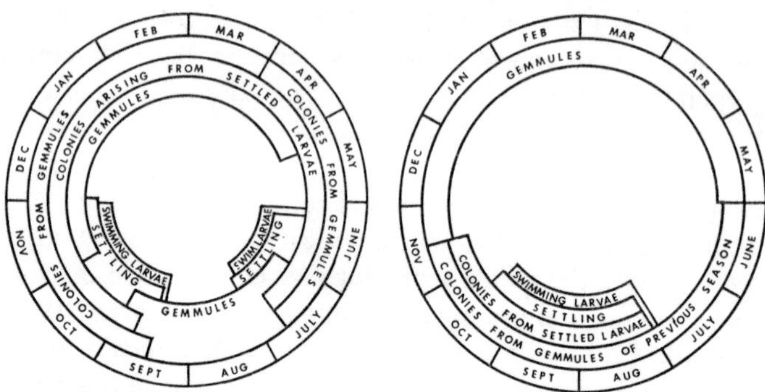

Fig. 4.18 Life cycles of *Haliclona loosanoffi* from Hatteras Harbor (North Carolina) at left and from Milford Harbor (Connecticut) at right.

The occurrence of gemmules in this intertidal species permits survival through high summer temperatures (*left*) as well as the freezing winter conditions encountered in northern habitats (*right*). (Redrawn from Wells, Wells and Gray 1964.)

Table 4.2 *The life histories of three hibernating sponges from Long Island Sound* (*data from Fell 1974*).

Month	Water temperature (°C) at low		*Haliclona loosanoffi** (occurrence of)			*Halichondria bowerbanki** (occurrence of)		*Microciona prolifera*† (occurrence of)	
			Adult sponges	Large oocytes and/or embryos	Gemmules	Large oocytes and/or embryos	Hibernating form of the sponge	Large oocytes and/or embryos	Hibernating form of the sponge
March	–	6.0	–	–	+	–	+	–	+
April	12.0	15.0	–	–	+	–	±	–	+
May	14.5	19.5	+	+	+	+	–	+	–
June	22.5	23.0	+	+	±	+	–	+	–
July	23.0	23.0	+	+	±	+	–	+	–
August	26.0	23.0	+	–	+	+	–	+	–
September	22.0	19.0	+	–	+	–	–		
October	19.0	8.5	±	–	+	–	–	–	–
November	13.0	4.5	±	–	+	–	–	–	+
December	1.5	1.0	–	–	+	–	+		
January	0	–1.5	–	–	+			–	+
February	0.5	0	–	–	+	–	+	–	+

*From the Mystic Estuary, Connecticut. Water temperature was measured once during the first half and once during the second half of each month from 1969 to 1970 (Fell, in preparation).

†From the Milford–New Haven area of Connecticut (after Simpson 1968).

5 Sponge classification

5.1 Introduction

It was indicated earlier that the difficulties which have been encountered in attempts to derive a phyletic classification of the Porifera have in large part been responsible for the rarity of studies on the biology of sponges. The classification of any group has a central role as a study in its own right, but it also provides an indispensable framework into which workers from many disciplines may integrate their observations and thus attempt comparative studies. Unstable classification introduces uncertainties into all aspects of the study of the group concerned.

The four classes of the Porifera have presented very different problems to the systematist. These will be reviewed briefly and followed by a discussion of the characters which have proved important in arriving at the present categories. The classification of recent members of each class will then be outlined. Such interpretation as can be attempted of phylogenetic relationships within the Porifera will be reserved until the fossil record of sponges has been discussed.

Between 1870 and 1928 many workers proposed classificatory systems of the Demospongiae stressing different aspects of skeletal morphology, and substantial elements of these systems are still incorporated in present schemes. The most comprehensive of the earlier classifications of the Demospongiae was that of Topsent (1928) on whose arrangement most modern workers have built. Another notable contribution was that of De Laubenfels (1936), who, in a single memoir, made an effort to diagnose and classify all the known sponge genera. De Laubenfels's scheme for the orders of Demospongiae rested heavily on that of Topsent, but subordinal arrangements and many generic diagnoses were his own. Despite inaccuracies, this memoir achieved something no other single work had done; it rendered the Demospongiae accessible as a group which could be studied using more modern methods. After 1936, it was possible to derive a picture of the range and structure represented in the group, and to think of refinements and improvements in classification. Previously this information was scattered through many works in several languages, and was disguised in different systematic schemes.

In a series of publications between 1953 and 1957, Levi proposed the first, and still the most important, reappraisal of the systematics of the Demospongiae, incorporating for the first time reproductive characteristics as parameters to define ordinal and subclass categories. Since that time, biochemical approaches interpreting higher level classification have been made (Bergquist and Hartman

1969; Bergquist and Hogg 1969) and other studies incorporating ecological data and electron microscopy are at the present time well advanced. With the relatively stable ordinal classification established by Topsent, Levi, and Bergquist and Hartman, and the suggestion by Levi of subclass groupings, traditional taxonomic studies since the early 1960s have made considerable progress toward assigning genera and species to their proper groups. There has been evidence of scientists building upon each other's work, in contrast to the earlier tendency, where each systematist evolved his own system, and rarely incorporated the ideas of others.

Other classes of the sponges have not proved as difficult to classify as the Demospongiae, which, after all, account for about 95 % of recent sponges. The Hexactinellida have received little attention since early this century, and no serious objections have been raised to the division of the recent Hexactinellida into two subclasses defined by microsclere type, established by Schulze in his later works. Palaeontologists find this an inadequate classification but, in the absence of further biological information, zoologists prefer to retain it.

Calcarea have been classified in two different ways. First, according to their level of organization as simple asconoid forms, the Homocoela, or more complex syconoid and leuconoid types, the Heterocoela. Secondly according to a complex of reproductive and histological characters into two groups, Calcinea and Calcaronea, both of which include simple asconoid types as well as more complex sponges. Both classifications date from last century, the former from Polejaeff (1883), and the latter from Bidder (1898). Recent publications by Hartman (1958*b*), Vacelet (1970) and Borojevic (1969) have favoured an updated version of Bidder's classification, although Tuzet, who has worked extensively on the Calcarea, holds to the older format (Tuzet 1973*a*). Despite the division of opinion, the fact that the Calcarea is a small class which has received consistent attention, particularly with regard to histology and development, has meant that generic and specific classification has been reasonably reliable. Reproductive information now exists for most groups of Calcarea and this renders Bidder's general scheme very usable. It is now considered that the Pharetronida rank with Calcinea and Calcaronea as subclasses of the Calcarea. The recent exciting discovery of living Sphinctozoa (Hartman, personal communication), a group thought to have been extinct since the end of the Cretaceous, raises the possibility that they too, when described fully, will prove to be distinct at a higher level rather than remain as an order of Pharetronida.

Last there is the Sclerospongiae, the recently proposed fourth class of the Porifera. Here problems centre not so much on the classification within the group, but on whether sclerosponges are distinct at the class level from Demospongiae. In upholding the view that they should be recognized as a class, the most cogent argument has been the long fossil history throughout which the two types of organization have remained distinct. The universal presence and complex structure of the massive calcareous skeleton also distinguish the Sclerospongiae from the Demospongiae.

5.2 How sponges are classified

5.2.1 Skeletal characteristics

Because, historically, the greatest numbers of species were found by expeditions and channelled to specialists, frequently after a lapse of many years, the classifications devised late last century, and up to 1936, drew upon descriptions of fixed or even dried material. This meant naturally that the primary feature for description was the skeleton of the sponge; then, if the organism had not been damaged during collection, the shape and surface morphology could also be considered. However, a systematist wishing to devise a classification to embrace all genera and species would have to work on skeletal characters; indeed they are the only content of many original descriptions.

Features of the skeleton which have been used frequently in classification are as follows:

(a) Chemical nature of the inorganic material

Whether siliceous or calcareous, and if the latter, whether the crystal structure is calcitic or aragonitic.

(b) Size categories of spicules

Discrete mineral skeletal elements, the spicules, occur in a range of sizes and within the Demospongiae and Hexactinellida both megasclere and microsclere categories are frequent. It is not possible to state the actual size limits which define microscleres as opposed to megascleres, but there is a clear functional difference. Megascleres are components of the primary skeleton, and their disposition is responsible for the form of the sponge and the development of regional internal substructure. Microscleres do not share this primary structural role, but are found as packing between megasclere tracts or scattered in surface or internal membranes.

In all groups it is possible to recognize regional differentiation in the skeleton. This becomes highly important in the Calcarea which lack microscleres and in the Hexactinellida where skeletal information provides almost the entire basis for classification. In the Demospongiae also, spicule type may vary according to location in the sponge; there may be a principal spicule type forming most of the skeleton, an accessory type either surrounding these or projecting from the principal tracts, and one or several categories of superficial spicules which may distinguish a cortical or ectosomal region. Microscleres can also be zoned in the sponge with distinct types reinforcing the superficial layers, and others lining deeper canals.

(c) Shape and size of the megascleres (Fig. 3.3, p. 92)

Spicules occur in many forms and some forms can be found in any group; for example, the simple diactine with pointed ends, an oxeote spicule, while others, such as the tetract, characterize certain orders and are thus an immediate aid to preliminary classification.

In addition to major structural differences which produce distinct spicule types there can be, within the Demospongiae, a clear size distinction between megascleres of identical construction which occur in different orders. For example, the structural oxeote spicules of Choristida and Spirophorida can be extremely large. Those of the Axinellida and Halichondria are smaller, and those of the Haplosclerida smaller again. In absolute terms, the range is from around 60 μm in the Haplosclerida to over 3000 μm in the Choristida, with 600 to 1200 μm being a common size in the Halichondria. Differences of this order are apparent immediately and are thus a useful systematic criterion.

Another feature which varies greatly is the degree of ornamentation on a spicule; this most frequently is the result of the presence or absence, and the quantity and size, of spines. While care must be taken to establish the limits of variation within specimens and species, there are still many cases where spicule ornamentation has proved useful in description.

When assessing the number of megasclere types which characterize a particular sponge, and when determining their location in the sponge, it is most important that the observer is not misled by the presence of foreign spicules. These occur frequently, sometimes as a result of selective accretion by sponges which elaborate their skeleton from detritus and loose spicules (as do many species of Demospongiae belonging to the orders Dictyoceratida, Haplosclerida and Poecilosclerida), and sometimes simply because they have settled on the sponge surface. Also, while the dominant megasclere type may be an oxeote, it is almost universal that occasional style and strongyle modifications will occur. Such variation at low incidence is trivial, but failure to realize this and to recognize foreign spicules has caused endless problems in identification.

There is a growing recognition that finer details of spicule length and thickness can vary with season of secretion and with geographic location within a species range. Thus any taxonomic decision which makes a species distinction on relatively minor spicule variation must be supported by statistical data. There are few instances in the literature where such distinctions have been made on an adequate basis.

(d) Shape and size of microscleres (Fig. 3.4, p. 93)

To a high degree microsclere types can characterize sponge genera. In the case of the Hexactinellida, the occurrence of amphidisc and hexaster microscleres is one of two characters which diagnose the subclass. Attempts to utilize microsclere type as a character to designate subclasses of Demospongiae have been misleading, and the old division into Astrotetraxonida and Sigmatotetraxonida developed on this basis cannot be upheld. In fact, only one order, the Spirophorida, is clearly defined by the presence of a particular type of microsclere, the sigmaspire, and even in that order some species may lack microscleres. At the generic and specific levels, microsclere type, and their size where one type occurs in several size categories, are essential taxonomic characters, particularly so in many poecilosclerid sponges.

(e) The molecular structure and extent of the organic skeleton

This has been applied as a systematic character only in the Demospongiae where the elaboration of a fibre skeleton frequently accompanies the organization of a spicule skeleton. There are also, in this group, species which lack a mineral skeleton. Some have spongin fibre skeletons, while others have reduced the skeleton to fundamentals and have only dispersed collagen fibrils remaining. Calcareous sponges do not have spongin fibre skeletons and in Hexactinellida, while spongin is present, it occurs only as a thin spicule sheath and as small patches of cement between adjacent spicules.

(f) The overall form and organization of the skeleton

In both the Hexactinellida and the Calcarea individual spicules, or spicules either interlocked or fused into networks, dictate the form of the sponge. Particular spicule types occur in specific locations and a highly specialized nomenclature accompanies this (Fig. 5.1). Massive secreted skeletons are common in some Calcarea and in the Sclerospongiae, and in these forms the living material is reduced to a superficial veneer. Form of the sponge is dictated by the skeleton. The same applies to one order of Demospongiae, the Lithistida, an artificial assemblage of species all of which have skeletons made up in large part of fused or interlocked spicules termed desmas (Fig. 5.12, p. 159). The form of this siliceous skeletal mass dictates the overall shape of the sponge.

In general, however, the overall form and arrangement of the sponge skeleton in Demospongiae is very flexible and the combinations of spongin and spicule, described in Chapter 1, produce certain easily recognizable skeletal arrangements. The pure spongin fibre skeleton of the Dictyoceratida is arranged as a tough anastomosing network (Fig. 5.21, p. 175), while that of the Dendroceratida branches from a basal point and ramifies without anastomosis (Fig. 5.23, p. 177). In the Haplosclerida an isodictyal arrangement is common in which spicules intersect according to a regular triangular or rectangular pattern and are surrounded by a variable quantity of spongin. Greater complexity is found in Poecilosclerida where isodictyal and also ramifying plumoreticulate forms occur. Here the principal spicules in the fibres are often surrounded by echinating accessory spicules which have their bases embedded in the fibres, and the fibres may be terminated at the surface by yet another type of spicule. Great descriptive potential lies in the precise skeletal arrangement of poecilosclerid sponges (Fig. 1.8, p. 46). Skeletal patterns of the above types produce sponges of very flexible growth form.

More rigid arrangements such as 'axial' patterns found in the Axinellida frequently produce a branching or lamellate sponge, with a condensed axis of spicule and fibre, around which cortical spicules are disposed at right angles (Fig. 5.16, p. 166). Superficial spicules of different types may be added.

Radial patterns (Figs. 1.7*e*, *h*, p. 44; 5.14, p. 164) depend on spicule rather than spicule fibre combinations and characterize the Spirophorida, Choristida and Hadromerida.

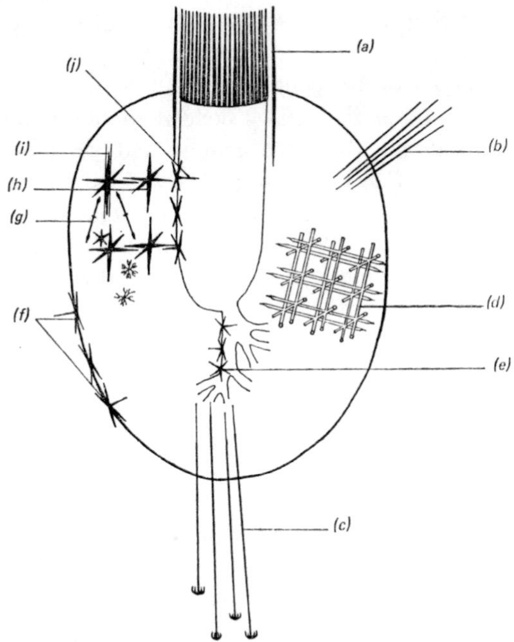

Fig. 5.1 Diagram of hexactinellid spicule arrangement.
(*a*) Prominent surface spicules, prostalia, are defined by their position. When fringing the osculum they are marginalia.
(*b*) Over the general body surface, the prostalia are termed pleuralia.
(*c*) Prostalia forming the rooting tuft are basalia.
(*d*) Dictyonalia, the parenchymal or principal body spicules which become fused to form the continuous skeletal framework of the Dictyonina.
(*e*) Canalaria. Spicules whose axial cross lies in the membrane which lines the exhalant canals.
(*f*) Dermalia. Spicules of the dermal skeleton.
(*g*) Parenchymalia. Intermedia. Body spicules situated between the principal skeletal elements.
(*h*) Parenchymalia. Principalia. Principal skeletal elements which can be free or fused.
(*i*) Parenchymalia. Comitalia. Thin body spicules situated very close to the principalia.
(*j*) Gastralia. Spicules which stand in definite relationship to the gastral membrane.
(Redrawn from Schulze 1887.)

As always, some groups may not be characterized except in a negative way. The Halichondrida have a skeleton termed 'halichondroid' which refers simply to a confused disposition of all but the dermal spicules.

5.2.2 External characters of the sponge

Shape of the sponge is often dictated by skeletal content and arrangement, and these features can characterize large groups and sometimes specify single genera (e.g. *Stylocordyla*, Fig. 1.7*f*, p. 44; *Thenea*, Fig. 5.9, p. 156). Frequently there is a certain polarity to the arrangement of inhalant and exhalant apertures, as for example in *Polymastia* (Pl. 9*b*) and *Latrunculia* (Fig. 5.15, p. 165), or many cup- or fan-shaped sponges where inhalant and exhalant faces are differentiated (Pl. 8*c*). These factors are diagnostic.

When gross sponge shape is used for taxonomic purposes it is essential that specimens be photographed where possible. The word amorphous, and other such nondescript terms, which so often spring to mind when looking at sponges, become tedious in works where many species are described.

5.2.3 Consistency of the sponge

One of the apparently subjective criteria used commonly in classification of Demospongiae is the consistency or feel of the sponge; whether fragile and easily torn, crumbly, brittle, tough and elastic, hard but breakable, stony, smooth or rough at the surface.

To a trained observer who understands thoroughly the types of skeleton which occur in sponges, these features of the whole organism convey very precise information regarding the type of skeletal elements present, whether spicule or fibre, and also the degree of their organization. For example, a disordered spicule skeleton, not welded into any form by spongin fibre, will crumble easily and such texture is typical of sponges belonging to the Halichondrida. Further, an anastomosing network of pure fibre will produce a sponge with soft texture, but one which is extremely tough, as for example in many Dictyoceratida. Sponges in which huge spicules constitute a high proportion of the body weight are often hard and certainly rough to the touch, as are all Hexactinellida, Spirophorida and Choristida.

It is also possible to detect by their consistency sponges which have high quantities of fibrillar collagen in the matrix. Members of the order Verongida and some Hadromerida have the firm, cartilaginous texture produced by emphasizing fibrillar collagen and de-emphasizing fibre and/or spicule skeletons.

Features unrelated to the skeleton which can yield significant information are as follows:

5.2.4 Pigmentation

Pigmentation of sponges is often held to vary widely within species and to be an unreliable systematic character. Certainly the aspect of the individual in relation to light has a great effect on the pigmentation of many sponges, most notably among the Demospongiae. It is extremely common to find that individuals in shade are non-pigmented, or weakly pigmented, while those exposed to light are fully pigmented. This applies also to shaded parts of an individual. It is most frequently noticeable in sponges where the normal pigmentation is black, grey, brown or purple. Most sponge pigments are carotenoids and in some micro-organisms these compounds have a photoprotective role. Solar radiation destroys the pigment before destroying metabolites which are vital to the animal (Litchfield 1976). Such a role for pigments in sponges, where colour could be interpreted as protection against light degradation of metabolites, is consistent with the ecological variability seen in sponge colouration.

Despite this variation, an intelligent observer who takes care to establish the range of colour in relation to light can use pigmentation with some certainty as a systematic character at the species level. It is frequent that there is a colour difference between the superficial and deeper regions of the body. This pattern is less likely to vary with microhabitat and is always a significant descriptive point.

5.2.5 Features of the sponge surface

In some groups the surface membrane of the sponge is smooth and supports a web of spicules (*Adocia*) or fibres (*Callyspongia*, Pl. 5*b*). In other cases it is merely reinforced by fibrillar collagen and mucoid material and tends to be elevated into projections at the points where it is intersected by spongin fibres or spicule tracts. The presence and size of these elevations or 'conules', and the disposition of inhalant apertures between them, are often characters of some use in generic or specific diagnosis.

In many species and genera, special surface structures are developed in relation to inhalant and exhalant apertures. All species of *Cinachyra* have sunken porocalyces, which are specialized inhalant and exhalant structures; all species of *Polymastia* have inhalant and exhalant siphons, and many Poecilosclerida have inhalant structures known as cribripores (e.g. *Hamigera*). Such features are diagnostic. However, number of oscules and pores and their precise size and arrangement in most sponges are a reflection of hydrodynamic factors and thus of little value in taxonomy. The way in which exhalant canals converge toward the oscules can have a significant effect on surface appearance. If many canals run at the surface toward a single osculum the stellate pattern produced is most marked, and is a useful descriptive point (Pl. 5*a*). The number of such oscules per unit area is not significant.

5.2.6 Reproductive characteristics

In Demospongiae and Calcarea sufficient information on larval morphology and mode of production is now available to enable these factors to be considered when confirming or reorganizing classification which is based on skeletal and gross morphological characteristics. No larvae are yet known for Sclerospongiae and only one incomplete reproductive sequence is known for the Hexactinellida.

In Calcarea the subdivision into subclasses is consistent with the occurrence of coeloblastula larvae in the Calcinea and amphiblastula larvae in the Calcaronea. The Pharetronida, in so far as their larvae are known, appear to have both types. This is consistent in a group which could be considered ancestral to the other two (Vacelet 1970).

It is in the Demospongiae that the new interpretation of classification which has proved possible as a result of amassing reproductive information has been most spectacular. All earlier classifications based high-level, subclass groupings on skeletal characters, and all had obvious inconsistencies. This had the effect, in the work of Topsent and De Laubenfels, that subclass groupings were not emphasized; primary importance was placed on defining orders, but not on relating them one to the other.

With Levi's (1956) suggestion that there were two subclasses within the Demospongiae (one the Ceractinomorpha in which sponges incubate parenchymella larvae, and the other the Tetractinomorpha in which sponges extrude gametes which develop externally to produce larvae of diverse type, some amphiblastulae and some peculiar parenchymellae), it was possible at last to recognize affinities between orders. The Ceractinomorpha is a group of six orders, five of which are obviously closely related. Detailed information on many Tetractinomorpha was still lacking in 1956 when this scheme was first proposed. There have been many subsequent alterations in the classification and composition of this subclass as more information has become available, culminating in Levi's (1973) suggestion that there are four subclasses, one of which, the Homosclerophorida, produces only amphiblastula larvae. The Tetractinomorpha, thus reduced, still show a variety of reproductive patterns which will be designated below for each order. The fourth subclass in Levi's classification is the Sclerospongiae, which is treated here as a separate class. Despite continued lack of information on some groups and the possibility of further modification, the impact of applying reproductive parameters to classification of sponges cannot be overestimated.

5.2.7 Biochemistry

Comparative biochemistry has been applied to sponge classification at frequent intervals, but serious attempts to combine competent taxonomy with competent chemistry have been rare. In the 1940s and 1950s Bergman, a respected natural-products chemist, combined his efforts with that of De Laubenfels, a sponge systematist, and many papers interpreting relationships within the Demospongiae in the light of comparative biochemistry particularly of sterols, resulted from their collaboration.

Of more recent date only Bergquist and Hogg (1969) and Bergquist and Hartman (1969) have combined comprehensive biochemical analyses with adequate taxonomy. In this area much effort has been and can be wasted by erroneous identification, but the potential value of biochemical characters in examining and establishing relationships is certain. However, systematists must do the work, ultimately the questions to be answered are biological. One can anticipate great activity in this area in the future for, provided the questions are posed on the basis of an understanding of established classification, presence or absence of particular molecules which can be identified and inter-related precisely provide objective characteristics which can indicate relationships. All work of this type must be introduced by screening as wide a range of species as possible in order to assess, first, the information content of particular molecular species in particular taxa, and second, the level of classification at which the information is most meaningful (see Chapter 7).

5.2.8 Histology

With the publication of a considerable number of papers dealing with electron microscopy of sponge tissue, it has become possible to categorize cell types with accuracy. Thus the occurrence of particular cell types, notably the cells with inclusions, can be used in taxonomic work.

Histological characters, particularly the position of the choanocyte nucleus, whether apical or basal, have long been used to distinguish the subclasses of the Calcarea. The arrangement of the choanocyte layer and the general shape of the choanocyte chambers have taxonomic significance in all groups. In Demospongiae the most common arrangement is into small spherical chambers. However, in the Dictyoceratida and Dendroceratida, two other arrangements are found: large oval sac-like chambers characterize the Dysideiidae and the Aplysillidae, while the Halisarcidae have long, narrow, branching chambers (Fig. 5.24, p. 177).

Despite a number of careful histological studies using light microscopy, particularly in the 1930s and 1940s, there was limited application of general comparative histology to sponge taxonomy until 1956 when Levi emphasized the importance of differences in certain cell types, between species of *Halisarca*. Since 1956, comparative histology has been used frequently in classification.

5.2.9 Ecology

The relevance of ecological study to taxonomic problems lies mainly at the species level. Here the problems of diagnosis can be acute and if differences in habitat preference, reproductive period, growth period or form can be demonstrated, then other morphological and biochemical differences, which may be very slight, can be better interpreted. Two papers dealing with species of *Halisarca* (Levi 1956; Chen 1976) provide excellent examples of systematic distinctions which can be based on ecological facts in combination with histological and reproductive information. In Demospongiae such as *Halisarca* where there is no mineral or fibre skeleton, and where the sponge itself is always a thin spreading mat, it is absolutely necessary to consider the biology and biochemistry of the sponge if one wishes to avoid the situation where all specimens would be relegated to a single species. A comparison between three *Halisarca* sp. is given in Table 5.1, and this illustrates the type of approach necessary to distinguish species in problem genera of which, regrettably, there are quite a number.

Table 5.1 *Biological differences among members of the genus* Halisarca *(after Chen 1976)*

Characteristic compared	*H. dujardini*	*H. metschnikovi*	*H. nahanthensis*
Habitat	Marine subtidal	Estuary	Marine intertidal
Spherulous cells	Equal cytoplasmic inclusions	With one large cytoplasmic inclusion	Equal cytoplasmic inclusions
Sexuality	Dioecious (?)	Monoecious	Monoecious
Spermatogenesis	Synchronous; long period	Asynchronous; short period	Asynchronous; short period
Sperm	Disc-shaped head	Unknown	Lemon-shaped head
Haploid chromosome number	Unknown	30	22
Egg size (diameter; μm)	90	120	120–130
Larva	Completely flagellated	Posterior pole not flagellated	Completely flagellated
Larval diameter (μm)	90	130	150
Larval cavity	Simple	Folded	Folded
Rhagon	Simple	Choanocyte chamber branching	Choanocyte chamber branching

5.3 Classification of recent Hexactinellida

Classification of the hexactinellid sponges at the subclass level is based substantially upon the form of the microsleres and the organization of the choanocyte layer. At the ordinal level, subdivision is on specific microsclere type coupled with the arrangement of the large hexactinal megascleres, whether separate or fused. The degree of fusion of the megascleres, however, is known to vary greatly with the age of the sponge and can be an unreliable guide to taxonomic affinity. In the absence of better criteria it is still used. Species and genera of Hexactinellida are defined largely upon the overall morphology of the sponge and upon the dimensions, type and precise localization of the spicules (Figs. 5.1, 5.2).

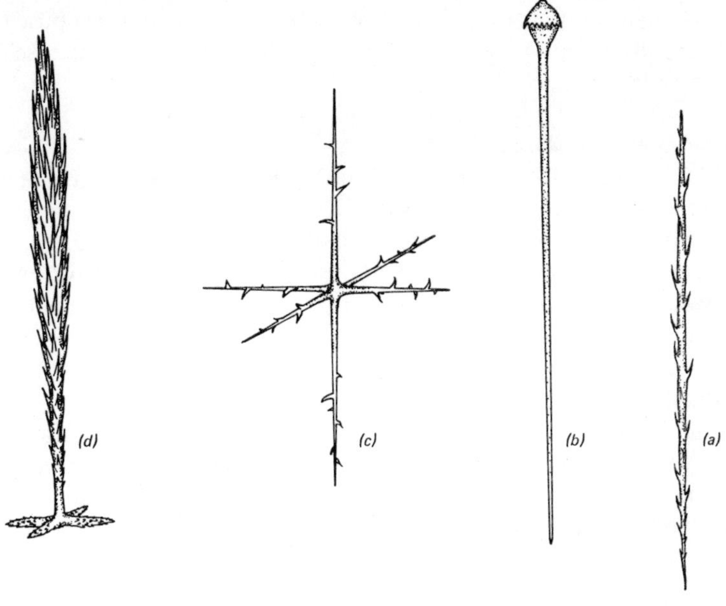

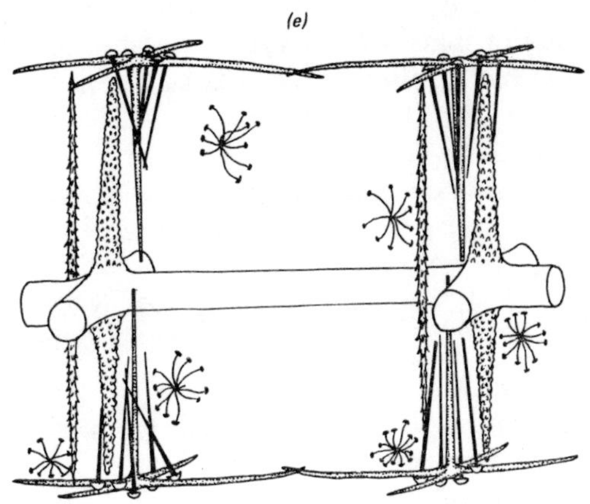

Fig 5.2 Hexactinellid spiculation.
(*a*) Uncinate (*Farrea*); (*b*) gastral clavula (*Farrea*); (*c*) oxyhexact with spined rays (*Semperella*); (*d*) autodermal pentact, a pinnulus (*Hyalonema*); (*e*) spicules in their natural position in a cross-section of the tube wall of *Farrea sollasii*. (All redrawn from Schulze 1887.)

Class **Hexactinellida** Schmidt

Exclusively marine sponges with a siliceous mineral skeleton composed largely of spicules with a hexactinal structure. Both megascleres and microscleres are always present.

Subclass **Amphidiscophora** Schulze

Microscleres are amphidiscs (Fig. 1. 5*b*, p. 24), hexaster microscleres are absent. The megasclere skeleton is composed of discrete spicules which never fuse to form a rigid network. Sponges are often of elaborate shape; they are never fixed firmly to a hard substratum, but live anchored into the sediment by a basal spicule tuft. Choanocyte chambers are disposed in irregular diverticula and are poorly set off from the reticular membrane.

Order **Amphidiscosa**

The amphidisc microscleres are always of birotulate form, symmetrical at both ends. All recent species fall into this order. Some fossil types have been classified in a second order, the Hemidiscosa, on the basis of having hemidisc microscleres in which the ends are asymmetrical.

Hyalonema (Fig. 1.5*a*, *b*, *e*, p. 24)

Subclass **Hexasterophora** Schulze

Hexactinellida with microscleres of the hexaster type (Fig. 1.5*f*). Megascleres of the sponge body are sometimes free, but more frequently are fused into a rigid skeletal framework (Fig. 5.2*e*). Hexasterophorans normally grow attached to a hard substrate, but it is not uncommon for them to be anchored by means of basal spicule tufts or mats. Choanocyte chambers are thimble-shaped and clearly set off from each other (Fig. 1.6*a*, *c*, p. 26).

Order **Lyssacina** Zittel

Hexactinellida in which the structural spicules either remain isolated, or become partly united in an irregular fashion bound together by siliceous cement or ladder-like trabeculae.

Euplectella (Fig. 1.5*d*, p. 24), *Lophocalyx* (Fig. 1.5*c*, p. 24).

Order **Dictyonina** Zittel

Hexactinellida in which the principal hexactinal spicules are united at an early stage into a compact framework in regular fashion.

Farrea (Fig. 5.2*a*, *b*, *e*), *Aphrocallistes*.

5.4 Classification of recent Calcarea

Classification of calcareous sponges at the subclass level hinges upon the particular combination of three major characters present in any sponge: first, the type of free larva produced, whether coeloblastula or amphiblastula; second, the position of the nucleus in the choanocyte, whether apical or basal in the cell body, and associated with this, how the flagellum arises from the choanocyte, either directly from the nucleus, or independently; lastly, particular attention is given to the presence or absence of triradiate spicules in the form of a tuning fork, in company with some type of calcareous carapace or fused skeleton. At the ordinal level, primary emphasis is placed on the degree of folding of the choanocyte layer, that is, whether the sponge is asconoid, syconoid or leuconoid.

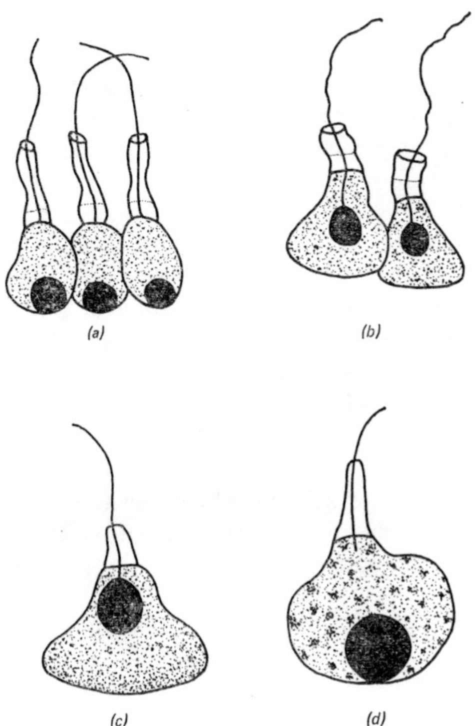

Fig. 5.3 Choanocytes of Calcarea.
(*a*) *Clathrina coriacea* (Calcinea) with basal nuclei and independent flagella.
(*b*) *Leucosolenia complicata* (Calcaronea) with apical nuclei from which the flagella arise.
(*c*) *Petrobiona massiliana* (Pharetronida) with apical nucleus from which the flagellum arises.
(*d*) *Murrayona phanolepis* (Pharetronida) with basal nucleus and independent flagellum.
(*a*, *b*, redrawn from Minchin 1900; *c*, *d*, redrawn from Vacelet 1961.)

Class **Calcarea** Bowerbank
Exclusively marine sponges in which the mineral skeleton is composed entirely of calcium carbonate. Skeletal elements are not differentiated into megascleres and microscleres.

Subclass **Calcinea** Bidder
Calcarea in which the free larvae are blastulae which can become solid by cellular ingression to produce a parenchymella-like structure, the coeloblastula (Fig. 4.9*i*, p. 115). Choanocyte nuclei are basal in the choanocyte and the flagellum arises independently of the choanocyte nucleus (Fig. 5.3*a*). Triactinal spicules when present are mainly equiangular (Fig. 5.5*b*).

Order **Clathrinida** Hartman
Calcinea in which the spongocoel is asconoid and lined by choanocytes throughout life. There can be some folding inwards which is not, however, accompanied by parallel folding of the pinacoderm and mesohyl. In these simple forms no cortical development is possible.

Clathrina.

Order **Leucettida** Hartman
Calcinea with syconoid to leuconoid construction of the canal system, thus having choanocytes restricted to choanocyte chambers. Elaboration of dermal and cortical structures is evident in all species.

Leucettusa (Pl. 8*b*), *Leucetta.*

Subclass **Calcaronea** Bidder
Calcarea in which the larvae are amphiblastulae (Fig. 5.4), the choanocyte nuclei are apical and each flagellum arises directly from the choanocyte nucleus (Fig. 5.3). Tractinal spicules are predominantly sagittal or inequiangular.

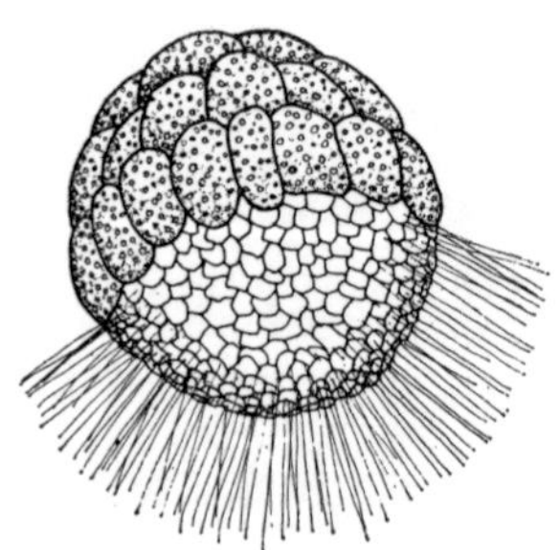

Fig. 5.4 Mature amphiblastula larva of *Sycon raphanus* (Calcaronea). (Redrawn from Schulze 1878.)

Order **Leucosolenida** Hartman

Calcaronea in which the spongocoel is lined throughout life by choanocytes, and no cortical development is seen.

Leucosolenia (Fig. 1.3*a*, *b*, p. 20).

Order **Sycettida** Bidder

Calcaronea with syconoid to leuconoid canal systems and choanocytes restricted to choanocyte chambers. Cortical and dermal structures are developed, and continuous, in all but one family.

Sycon, Aphroceras (Fig. 5.5*a*).

Subclass **Pharetronida** Zittel

Calcarea in which the calcareous skeleton is deployed in several distinct ways, as discrete spicules, spicules aligned in tracts, spicules welded into a framework or as an external or internal aspicular calcareous matrix. The cortical region is

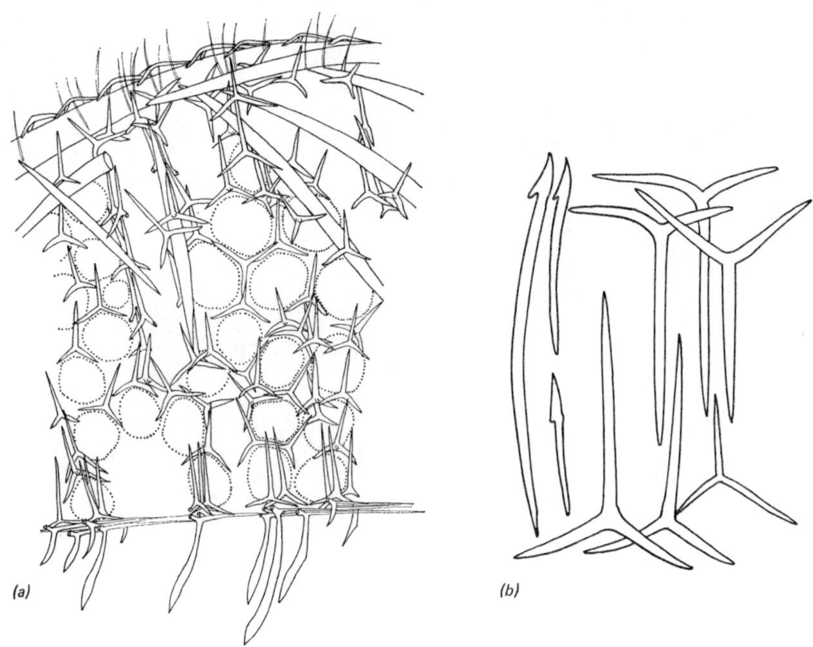

Fig. 5.5
(*a*) A radial section through an oscular tube of *Aphroceras ensata*, a leuconoid member of the Calcaronea. The zonation of different spicule types at the surface, through the choanoderm and matrix, and bordering the atrial cavity is shown. (Redrawn from Borojevic 1970*b*.)
(*b*) A selection of spicules from a calcareous sponge belonging to the Calcinea.

frequently armoured. Peculiar tuning-fork triactines are usually present, sometimes organized into tracts. An oscular armament of tetractinal spicules is present. Larvae and choanocytes of both calcinean and calcaronean types are found (Fig. 5.3*c*, *d*).

Order **Inozoa** Steinmann

Pharetronida with no segmented or compartmented structure and a canal system which is probably always leuconoid.

Paramurrayona (Fig. 5.6*a*, *b*), *Petrobiona, Murrayona* (Fig. 5.6*c*).

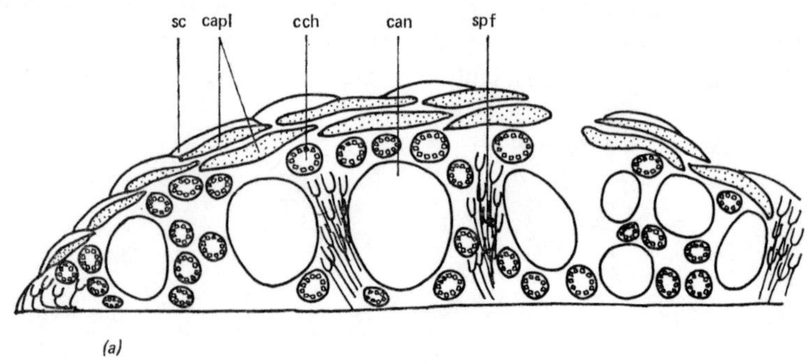

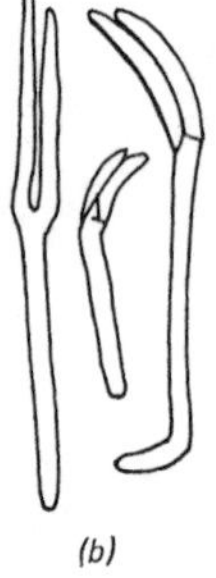

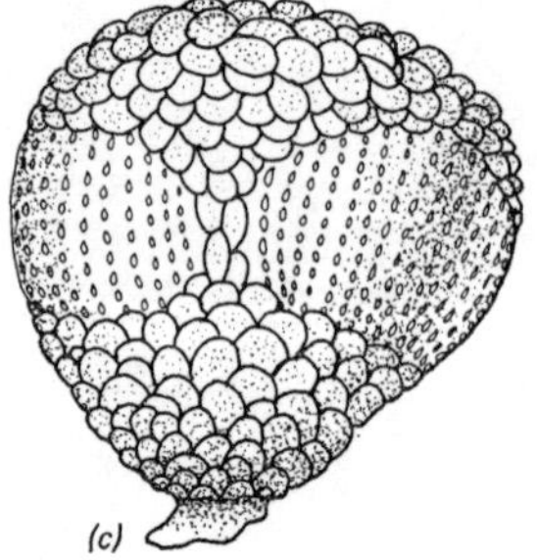

Fig. 5.6

(*a*) Diagrammatic vertical section of *Paramurrayona corticata* (Pharetronida). can, canals; cch, choanocyte chambers; spf, spicules forming fibres (tuning fork types); ca pl, calcareous plaques; sc, superficial scales.

(*b*) Tuning fork triradiate spicules of *Paramurrayona.*

(*c*) An individual of *Murrayona* showing the arrangement of the pore areas and the area covered by calcareous scales.

(*a* and *b*, redrawn from Vacelet 1964; *c*, redrawn from Kirkpatrick 1910.)

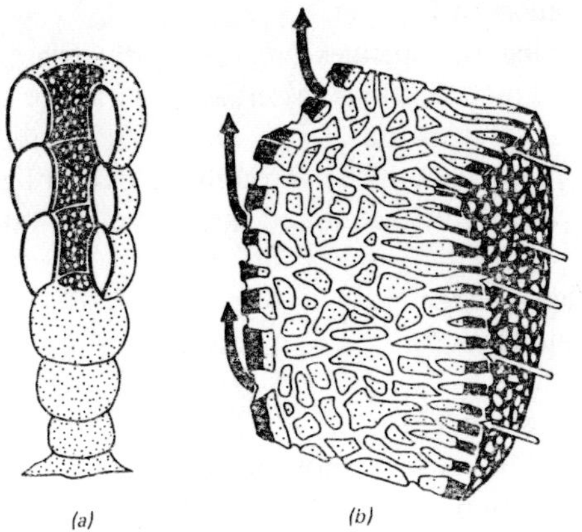

Fig. 5.7 The organization of sphinctozoans.
(*a*) A Triassic form, *Polytholoisa*, with the animal partly opened to show the central cavity.
(*b*) The channel system of *Polytholoisa* showing the direction of water circulation. (Redrawn from Ziegler and Rietschel 1970.)

Order **Sphinctozoa** Steinmann
Segmented Pharetronida with a rigid external skeleton. Details of the canal system are unknown, but it was probably of syconoid type. This group was until recently thought to be extinct, but living specimens have been found. No names are yet available for recent genera.

Representative fossil genera are *Sollasia, Polytholoisa* (Fig. 5.7).

5.5 Classification of recent Demospongiae

At the subclass level primary emphasis is given to reproductive pattern, whether oviparous or viviparous, and to the type of larva produced. Orders are defined on the type of megascleres and microscleres present, on the organization and composition of the skeleton and on the detail of reproductive patterns. At the family level in this diverse group, a variety of characteristics must be employed.

Class **Demospongiae** Sollas
Marine or fresh-water sponges with a siliceous skeleton in which megascleres are usually either monaxons or tetraxons, a triaxon being present as a major spicule type in one subclass only. There are microscleres of diverse types. The spicule skeleton can be supplemented or replaced by a spongin skeleton which is utilized either as a cementing element for the mineral skeleton, or to form fibres. Some genera have lost all specialized skeletal components.

Subclass **Homoscleromorpha** Levi
Demospongiae in which the spicules are frequently triactines with abundant diactine and tetractine modifications. Microscleres are not clearly distinguished from megascleres. All spicules are of very small size, usually less than 100 μm and are distributed in great numbers throughout the sponges with no regional organization. Embryos are incubated and the larvae are amphiblastulae.

Order **Homosclerophorida** Dendy
Diagnosis as for the subclass.

Family **Oscarellidae** Lendenfeld
Homosclerophorida which lack mineral or fibre skeleton, but which retain the pattern of larval production typical for the order and incubate amphiblastulae.
Oscarella.

Family **Plakinidae** Schulze
Homosclerophorida which retain the mineral skeleton and show a variety of spicule forms based on the di-, tri- and tetractinal pattern. The sponge body is never very complex.
Plakina (Fig. 1.7*a*, p. 44), *Plakortis* (Fig. 1.7*b*).

Subclass **Tetractinomorpha** Levi
Demospongiae in which the typical reproductive pattern is oviparous. The eggs are extruded and development, which is either direct or by way of a larval stage, is external. Incubation of miniature adults with deletion of a larval stage is found in one order. The typical larvae are parenchymellae, although it is clear that in the Hadromerida the strange blastula larvae of some genera such as *Cliona* and *Polymastia* will in future merit description as distinct types.

Megascleres are tetraxonid and monaxonid, together or separately, organized usually with a recognizable pattern which is either radial or axial. Microscleres are most frequently asterose (Pl. 6*a*) but sigmas and raphides also occur.

While the five orders united under this subclass do indeed show features in common, it is certain that the group is polyphyletic. What remains uncertain is how to group the orders, particularly in view of the incompleteness of our knowledge, in particular of the Choristida and the Axinellida.

Order **Choristida** Sollas
Tetractinomorpha in which the microscleres are asters, sometimes accompanied by micro-oxeas. The megascleres are tetractinal (Fig. 5.8) and oxeote, and always show some radial arrangement, which is most apparent at the sponge surface (Fig. 1.7*c*, p. 44). In large sponges the spicule orientation becomes confused towards the centre of the body. In some families a series of genera can be recognized in which either tetraxons, or microscleres, or both spicule categories can be lost, leaving only oxeote spicules. Radial skeletal organization remains

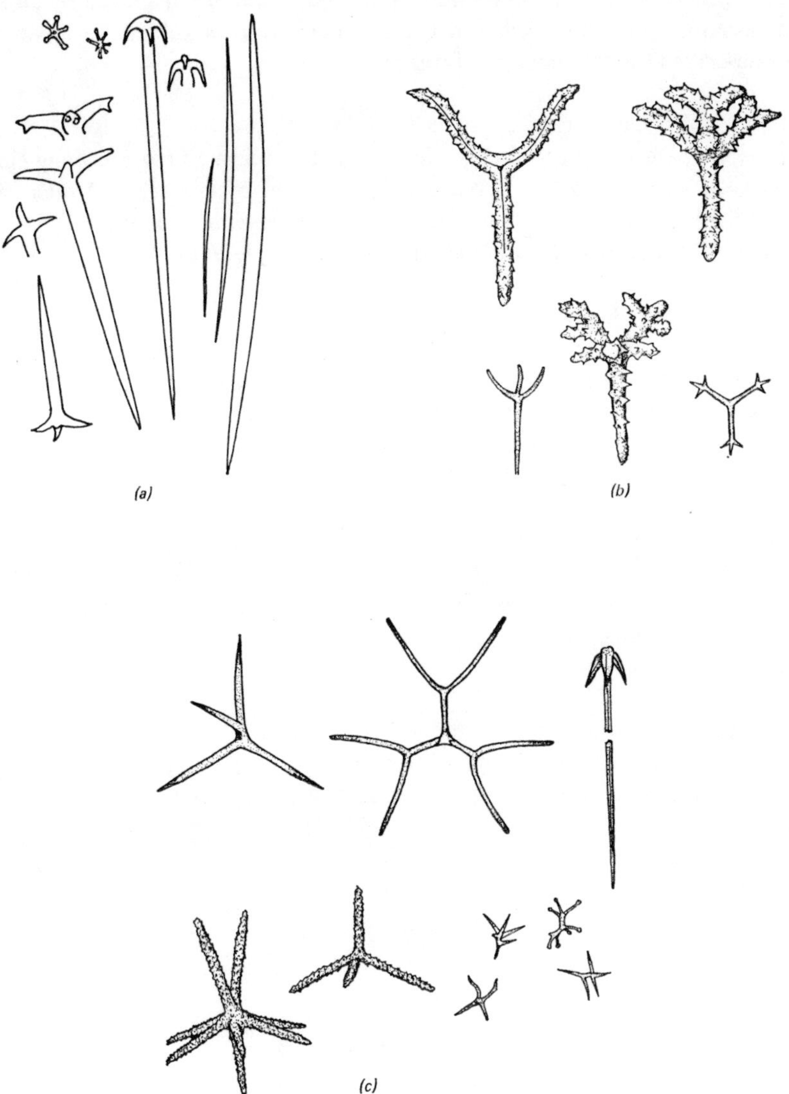

Fig. 5.8 Spicule complements of representative Choristida (Tetractinomorpha).
(*a*) *Stelletta durissima* (Stellettidae). Plagiotriaenes, dichotriaenes, anatriaenes, oxeas and tylasters.
(*b*) *Thrombus challengeri* (Thrombidae). Spined orthotriaenes, dichotriaenes and amphiasters.
(*c*) *Thenea novaezealandiae* (Theneidae). Dichotriaene, anatriaene, smooth and spined plesiasters, small streptasters.

in such genera and this, in addition to the coarse texture displayed by all forms emphasizing a siliceous skeleton, makes recognition possible. No larvae have been described for any member of this order.

Family **Stellettidae** Carter

This is the most typical family, in which the megascleres are long-shafted triaenes of various types and oxeas, accompanied by euaster and microrhabd microscleres. Sterrasters and amphiaster microscleres are absent.

Stelletta (Fig. 5.8*a*), *Ancorina, Disyringia* (Fig. 1.7*g*, *h*, p. 44).

Family **Geodiidae** Gray

Choristida possessing long-shafted triaenes and oxeas as megascleres and characterized by the presence of sterrasters (Fig. 3.4*c*, p. 93) among the microscleres. Typically these form a superficial cortical armour.

Geodia (Fig. 1.7*c*, p. 44), *Rhabdastrella*.

Family **Calthropellidae** Lendenfeld

Choristida in which some or all of the triaenes are of a specialized form, the calthrops (Fig. 3.3*i*, p. 92; Pl. 6*b*), in which no ray is markedly longer than the other three. Such spicules are frequently malformed or reduced to diactines. Microscleres are euasters.

Calthropella.

Family **Pachastrellidae**

Choristida possessing calthrops. Megascleres accompanied by microscleres which are streptasters of various types. Euasters are absent.

Pachastrella.

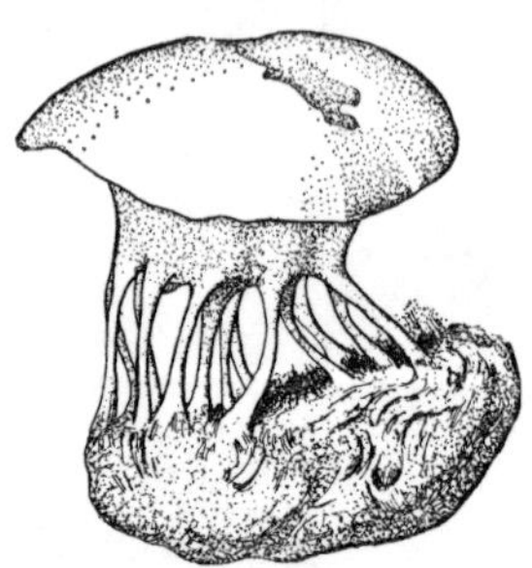

Fig. 5.9 Thenea wyvilli (Choristida: Theneidae).
As in many deep-water sponges the form of the body achieves some external symmetry. The members of this family characteristically possess rooting processes and an expanded upper oscular surface. (Redrawn from Sollas 1888.)

Family **Theneidae** Sollas

Choristida possessing normal long-shafted triaenes accompanied by streptasters of various forms. Euasters are lacking.

Thenea (Figs. 5.8*c*, 5.9).

Family **Thrombidae** Sollas

Choristida with megascleres which are exclusively small, spined, triaenes. Microscleres are amphiasters.

Thrombus (Fig. 5.8*b*).

Family **Jaspidae** De Laubenfels

Choristida which lack triaenes and have only oxeas as megascleres. These are disposed radially only at the surface. Microscleres are euasters, micro-oxeas or sanidasters, a special type of streptaster with straight axis and multiple rays at each end.

Asteropus, Jaspis.

Order **Spirophorida** Levi

Tetractinomorpha with a perfect radial skeleton and consequent near spherical form (Fig. 5.10). Megascleres are triaenes and oxeas, with the protriaene being a common and characteristic spicule at the sponge surface. The microscleres are

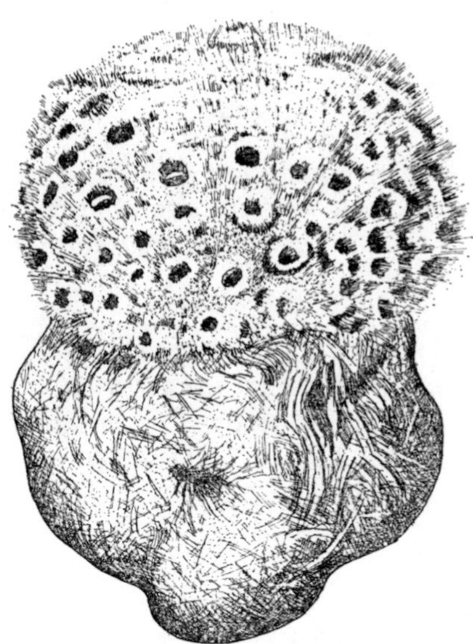

Fig. 5.10 Cinachyra barbata (Spirophorida).
An unusual large rooting tuft of spicules characterizes this deep-water, soft-bottom-inhabiting sponge. The specialized, inhalant and exhalant areas, the porocalyces, are distributed over the shaggy free surface. (Redrawn from Sollas 1888.)

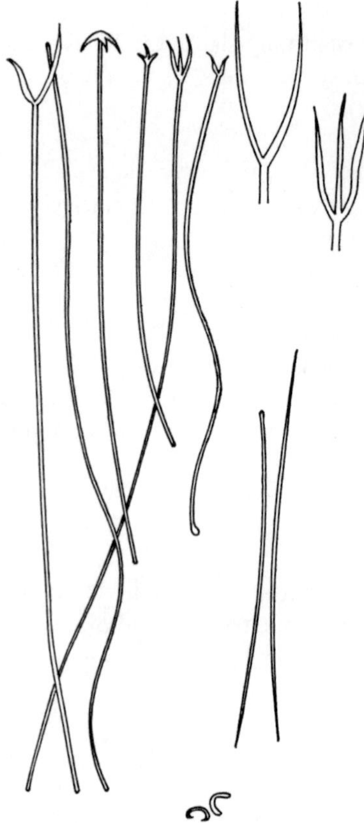

Fig. 5.11 Spiculation of a typical member of the Spirophorida, *Cinachyra vacciniata*. Protriaenes, anatriaenes, oxeas and sigmaspire microscleres. For detail of the sigmaspire structure refer to Fig. 3.4*j*. (redrawn from Levi 1973.)

unique sharply contorted sigmoid forms, termed sigmaspirae (Figs. 3.4*j* (p. 93), 5.11). Reproductive patterns range from the extrusion of fertilized eggs which fix to the substrate and develop directly, to incubation of complete young sponges which are then expelled by localized breakdown of the pinacoderm. No free larvae have been described.

Family **Tetillidae** Sollas
Diagnosis as for the order.
Tetilla, Cinachyra (Figs. 5.10, 5.11).

Order **Lithistida** Schmidt
Demospongiae of diverse origin which have developed a spicule type known as a desma. This can be either monaxonid (monocrepid) or tetraxonid (tetracrepid)

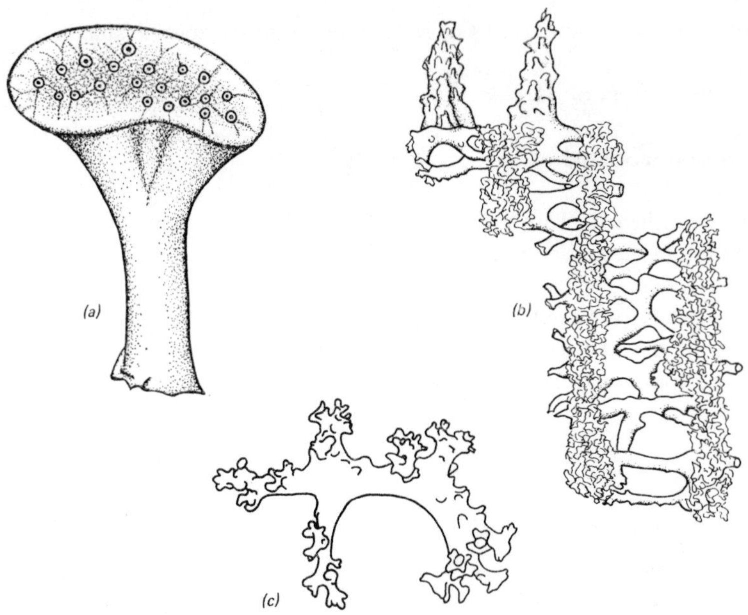

Fig. 5.12 Lithistid structure and spiculation.
(*a*) An individual of *Corallistes typus*; (*b*) section of the skeleton of *Corallistes typus* at right angles to the surface illustrating the fusion of the tuberculate desmas into a rigid structure; (*c*) an individual desma.
(*a* and *b*, redrawn from Schmidt 1870.)

in derivation, but always has complex branching pattern and these branches interlock to form an extremely hard stony skeleton (Fig. 5.12).

Other megascleres, and various microscleres, accompany the desmas and it is largely on the basis of accessory spicule composition, and disposition, that attempts have been made to assign lithistid genera to other orders of the Demospongiae.

No complete recent re-study of this group has been made, and attempts to divide the group are premature at this time. For the most recent subdivision of the Lithistida the reader is referred to Levi (1973). Family diagnoses will not be given here, but they are based entirely on skeletal characters.

Because of the interlocked desma skeleton, lithistid sponges are very common as fossils and frequently preserve their overall shape.

Order **Hadromerida** Topsent

Hadromerid sponges have megasclere skeletons composed of monactinal spicules which are always tylostyles or subtylostyles organized on a radial

Plate 7
Scanning electron micrographs of some representative sponge spicules.

(a) An anisochela of *Mycale*.

(b) An isodiscorhabd of *Sigmoscepterella*.

(c) An acanthoxea of an unidentified myxillid. The delicate terminal spining of some tylote megascleres is also shown.

(d) A representative spicule array for members of the family Clathriidae, including large and small styles (the larger ones spined), palmate isochelae and toxas. The spined strongyle to the right is a single aberrant spicule.

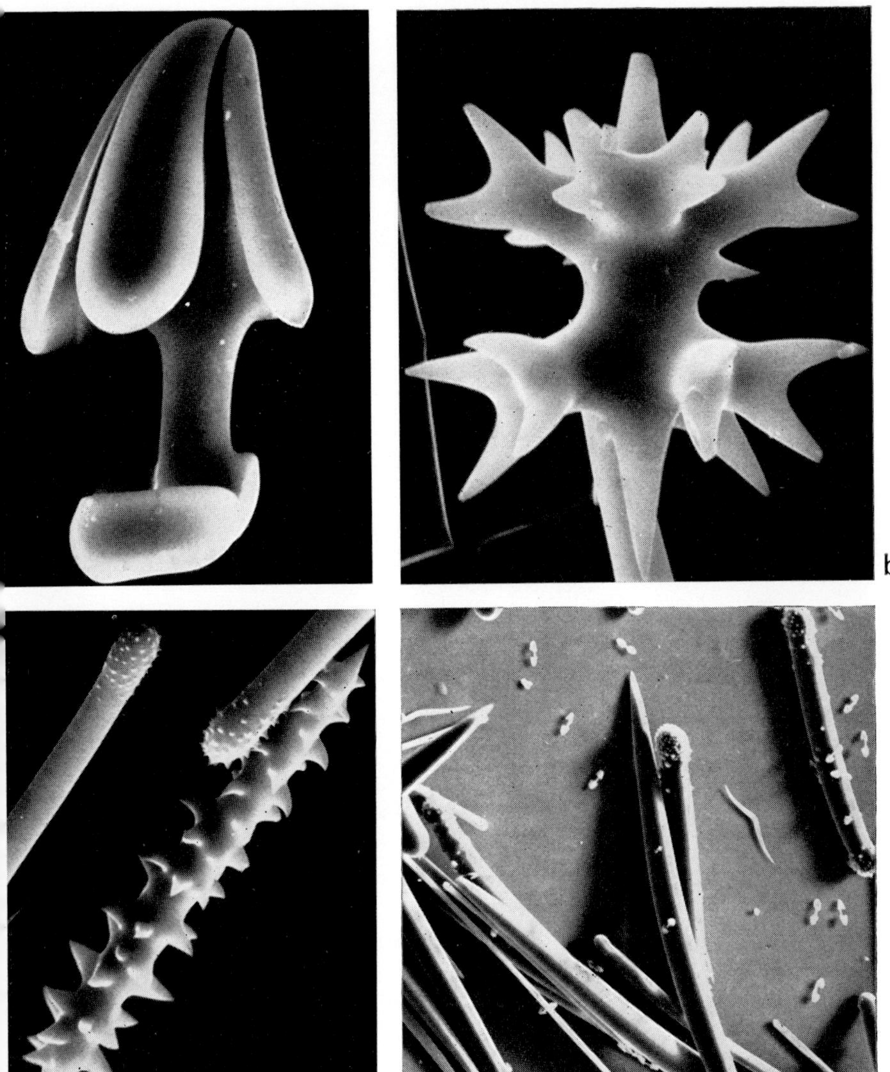
b
d

a

c

Plate 8

(a) *Iophon laevistylus* (Demospongiae: Poecilosclerida). A good example of a sponge with erect, branching tubular construction. Note the marked development of the oscular membrane to form a contractile diaphragm. (Photo R. Grace.)

(b) *Leucettusa lancifer* (Calcarea: Calcinea). A classic example of the compound vase construction of many Calcarea. (Photo W. Doak.)

(c) *Gellius imperialis* (Demospongiae: Haplosclerida). A huge cup sponge where the inhalant (outer) and exhalant (inner) surfaces are differentiated. (Photo W. Doak.)

pattern. This radial construction can be emphasized through the entire body (Tethyiidae) or evident only at the surface (Polymastiidae). There is usually a size distinction in the megascleres, with smaller spicules assuming a radial disposition at the surface and protruding some distance beyond the surface as a hispid fur.

Spongin is present in many Hadromerida, but never as fibres, thus the consistency of these sponges is firm but non-elastic. Microscleres are frequently absent in Hadromerida, but where present they are asterose or micro-oxeas.

Development of Hadromerid sponges, wherever it has been observed, is oviparous with eggs extruded and development external. However, the type of larva produced varies from a strange blastula (which is not to be equated with the amphiblastula of the Homosclerophorida) in *Polymastia* to a typical parenchymella in *Tethya*. Very few larval types have yet been described.

Family **Suberitidae** Schmidt

Megascleres are tylostyles, subtylostyles, or rarely styles; microscleres are most frequently absent, but microstrongyles can be present. The radial arrangement of the skeleton is evident only at the surface. The spicules have confused orientation in the deeper regions of the sponge, but in a few cases may assume a loose axial orientation.

Suberites.

Family **Polymastiidae** Gray

Megascleres are tylostyles or subtylostyles, always divisible into two or three clear size categories. Microscleres are rare, but acanthose micro-oxeas can be present. The massive or spreading body with erect oscular and pore-bearing papillae is characteristic for the family.

Polymastia (Pl. 9*b*).

Family **Spirastrellidae** Ridley and Dendy

Encrusting or massive erect Hadromerida in which the megascleres are tylostyles and the microscleres spirasters (Fig. 5.13*b*). These sponges are very similar in morphology to the Clionidae, but only in rare cases have they been demonstrated to bore into calcareous material.

Spirastrella.

Family **Clionidae** Gray

Hadromerid sponges which excavate burrows in any calcareous material, coralline algae, shells or coral. Megascleres are tylostyles and the microscleres, spirasters, micro-oxeas or amphiasters. Inhalant and exhalant papillae are always present.

Cliona (Pl. 10*b*; Fig. 5.13*b*), *Alectona* (Fig. 4.16, p. 130).

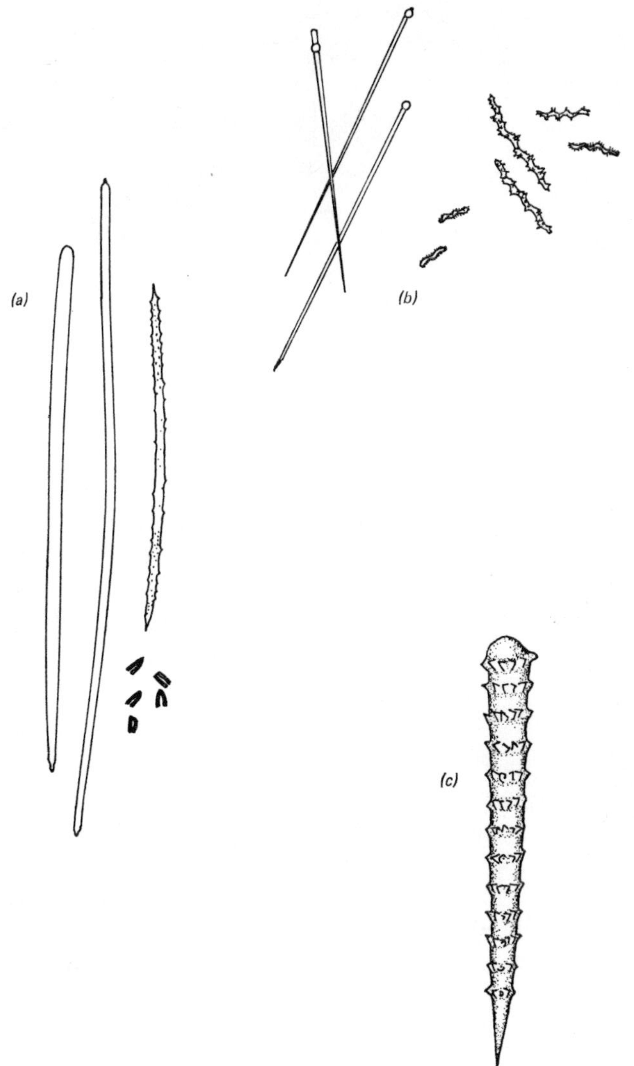

Fig. 5.13 Spicule complements of representative Hadromerida and Axinellida (Tetractinomorpha).
(*a*) *Myrmekioderma granulata* (Axinellida, Desmoxyidae) styles, oxeas, acanthoxeas and microscleres which are raphides aggregated into small, tight groups.
(*b*) *Cliona lobata* (Hadromerida, Clionidae) tylostyles and spirasters.
(*c*) *Agelas mauritiana* (Axinellida, Agelasidae) acanthostyle, spined in a verticillate pattern.

Fig. 5.14 Section of *Tethya* at right angles to the surface. The strong tissue cortex is reinforced by cortical megascleres in brushes, and dispersed microscleres (spherasters). The radiating tracts of endosmal megascleres grade into the cortical brushes. *Tethya* produces stalked surface buds as one means of asexual reproduction. (Redrawn from Brien 1973.)

Family **Tethyidae** Gray

Massive, usually spherical Hadromerida with a perfect radial megasclere skeleton and a marked cortical region (Fig. 1.7*d*, *e*, p. 44). Megascleres are never tylostyles, but are a special type of subtylostyle known as a strongyloxea. This is a spicule in which the two ends are different, usually rounded at the broad end and tapering to a point distally. Both ends are more asymmetric than those of typical styles and the spicule diameter is frequently increased or decreased in abrupt steps. Microscleres, where present, are spherasters and micrasters.

Tethya (Pls. 6*a*, 10*a*; Figs. 1.7*d*, *e* (p. 44) 5.14), *Aaptos*.

Family **Chondrosiidae** Schulze

Massive or encrusting Hadromerida in which the megasclere skeleton has been lost and in which the microscleres, when present, are euasters. Some forms lack spicules entirely. The surface is often smooth and a marked cortex, enriched strongly with fibrillar collagen, is present. This family is placed in the Hadromerida largely on the evidence of biochemical studies and microsclere morphology.

Chondrilla, Chondrosia.

Family **Stylocordylidae** Topsent
Hadromerida with asymmetrical globular or ovoid body supported on a long stalk. These sponges are inhabitants of deep water and can grow on soft bottoms, where their rooting spicules bind fine debris for attachment. The megascleres are styles or oxeas arranged according to a radial pattern, converging towards the stalk in which the spicules are disposed along the axis. Microscleres, when present, are micro-oxeas, microstrongyles or asters.

Stylocordyla (Fig. 1.7*f*, p. 44).

Family **Placospongiidae** Gray
Hadromerida with tylostylote megascleres and sterrasters as microscleres. The latter are arranged as a hard cortical crust subdivided into a polygonal pattern by pore grooves.

Placospongia.

Family **Timeidae** Topsent
Encrusting Hadromerida possessing tylostyles and euasters.

Timea.

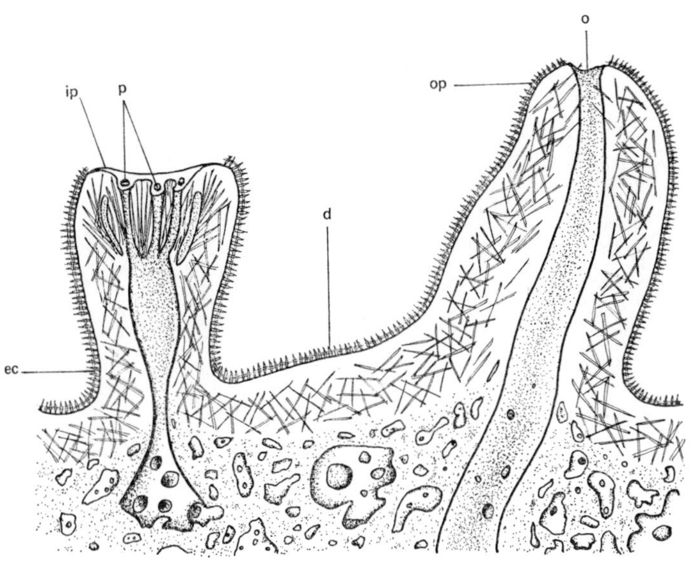

Fig. 5.15 A vertical section of the superficial region of *Latrunculia apicalis* (Hadromerida). The section is taken through an inhalant papilla (ip) and a oscular papilla (op). Such specialized surface structures are not uncommon in Tetractinomorpha. The alignment of the discorhabd microscleres (d) in a surface layer and the specialization of an ectosomal region (ec) set off from the underlying choanosome are evident: p, pores; o, osculum. (Redrawn from Ridley and Dendy 1887.)

Family **Latrunculiidae** Topsent

Hadromerida possessing discorhabd microscleres (Fig. 3.4*f*, p. 93), often aggregated into a dense dermal crust. Megascleres are styles, radially disposed at the surface, with axial orientation in stalked forms and confused disposition in the choanosome of massive forms. Incubated young sponges are sometimes found in this usually deep-water group.

Latrunculia (Fig. 5.15), *Sigmoscepterella* (Pl. 7*b*).

Order **Axinellida** Bergquist

Tetractinomorpha with a spicule and fibre skeleton condensed into an axis from which diverges a plumose or plumoreticulate extra-axial skeleton which can be reinforced strongly with spongin fibre. The megascleres are monaxons, oxeas, styles or strongyles in all combinations. The spicules are often sinuous, curved or irregular at one end. A stiff axial region, distinct from a softer, extra-axial region, is typical and, although massive and encrusting forms are found, these always preserve an axial component, either parellel to the substrate (*Bubaris*) or within the small processes which make up the massive body (*Pseudaxinella*). The sponge surface is usually rough with projecting spicules; few, if any species, have smooth surfaces.

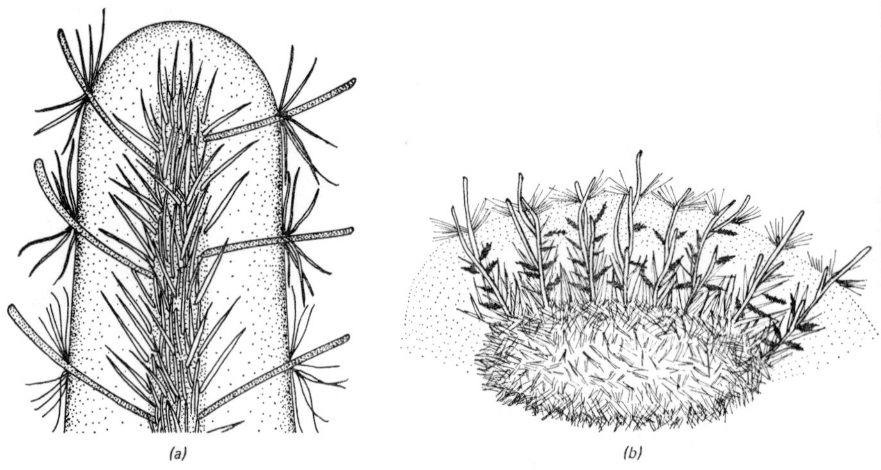

Fig. 5.16 Skeletal organization of Axinellida.

(*a*) Vertical section through the thin lamellate body of *Ceratopsion*, a genus with a simple axinellid skeleton, incorporating a condensed axis with extra-axial megascleres diverging from this. Where the long extra-axial spicules pass through the sponge surface, they are surrounded by brushes of dermal spicules.

(*b*) Diagrammatic transverse section of a branch of *Raspailia compressa*. A dense spicule and fibre axis supports extra-axial spicule tracts which in turn support echinating accessory spicules. A special dermal skeleton is present surrounding all points where the extra-axial tracts pierce the surface membrane.

Zonation of the various spicule categories in the sponge aids greatly in the identification of Axinellida. Very complex skeletons with distinct axial, extra-axial, echinating and ectosomal spicules are common (Fig. 5.16*a*, *b*).

Microscleres are often absent, but particular types categorize some of the families; raphides and microxeas are the most common, asterose and sigmoid forms occur as do several quite unique forms.

In the few species where reproductive processes have been observed, axinellids are oviparous and the larvae are parenchymellae.

Family **Axinellidae** Ridley and Dendy
Axinellida in which microscleres other than raphides are absent.
Axinella (Pl. 9*a*), *Ceratopsion* (Fig. 5.16*a*).

Family **Trachycladidae** Hallmann
Axinellida with microscleres of extremely contort spiraster type, sometimes with the addition of short, strongylote microscleres termed rhabds.
Trachycladus.

Family **Desmoxyiidae** Hallmann
Axinellida with microscleres in the form of smooth or spined microxeas; other more complex types may accompany these.
Desmoxya, Myrmekioderma (Fig. 5.13*a*).

Family **Hemiasterellidae** Lendenfeld
Axinellida with euasters as microscleres.
Hemiasterella.

Family **Sigmaxinellidae** Hallmann
Axinellida with sigmoid microscleres.
Sigmaxinella.

Family **Raspailiidae** Hentschel
Axinellida in which the typical skeleton is supplemented by dermal spicule brushes, usually of fine styles grouped around a long central style, and by echinating acanthostyles or rhabdostyles, The latter may be very small, but true microscleres are absent. Sponges are erect and branching, encrusting or very rarely massive.
Raspailia (Fig. 5.16*b*), *Eurypon.*

Family **Rhabderemiidae** Topsent
Axinellida with principal skeleton made up of columns of rhabdostyles, the microscleres are extremely contorted sigmas, and thraustoxeas which are peculiar asymmetrical microxeas.
Rhabderemia.

Family **Agelasidae** Verrill

An enigmatic monogeneric family, with a reticulate fibrous skeleton in which the spongin fibres contain no primary coring spicules, but are echinated by unique acanthostyles with verticillate spines (Fig. 5.13*c*).

The form of the sponge body is variable, ramose, lamellate, tubular or massive, the colour is frequently orange or red and the texture like that of *Spongia* itself, extremely tough, but compressible.

In placing the Agelasidae in the Axinellida, emphasis has been put on biochemical characteristics, namely free amino acid patterns (Bergquist and Hartman 1969), sterol composition (Bergquist and Hofheinz, in press) and on the fact that two species of *Agelas* are known to be oviparous (Reiswig 1976). It is suggested that the Agelasidae are axinellids in which the condensed axial skeleton has been reduced completely and the reticulate, extra-axial elements emphasized to the extent that they make up the entire skeleton.

Subclass **Ceractinomorpha** Levi

Demospongiae in which the typical reproductive pattern is viviparous; the many species for which reproductive sequences are known all incubate parenchymella larvae. The megascleres in the group are always monaxonid, triaenes are never present. Microscleres are generally sigmoid or chelate (Pls. 6*c*, *d*, 7*a*), never asterose. Spongin is an almost universal component of the skeleton, being absent only in one family, the Halisarcidae.

Order **Halichondrida** Topsent

Ceractinomorpha in which microscleres are absent and megascleres are oxeas, styles or strongyles in many combinations. The skeleton shows no organization except at the surface, where a layer of tangential dermal spicules (Fig. 5.17), sometimes supported by ectosomal spicule brushes, can be organized. The endosomal skeleton with spicules in confusion is termed 'halichondroid'.

Parenchymella larvae are ciliated over their entire surface (Pl. 12*a*).

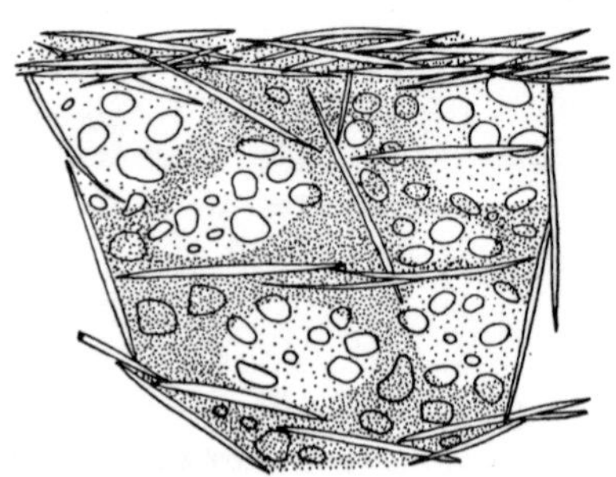

Fig. 5.17 Surface view of the tangential dermal skeleton of *Halichondria* showing inhalant pores and oriented spicules. (Redrawn from Hartman 1958*a*.)

Family **Halichondriidae** Vosmaer

The principal megascleres are diactines, mainly oxeas, although accessory styles may be present. A marked system of subdermal spaces is present and sets the dermal skeleton off clearly from the confused endosome.

Halichondria (Fig. 5.17; Pl. 12*a*), *Trachyopsis.*

Family **Hymeniacidonidae** De Laubenfels

The principal megascleres are styles, with occasional accessory oxeas. Hymenacidonid sponges frequently show surface and ectosomal skeletal organization, with spicule brushes that protrude beyond the sponge surface, but the endosomal skeleton has no orientation.

Hymeniacidon.

Order **Poecilosclerida** Topsent

This is the largest, and structurally most diverse order of the Demospongiae. Poecilosclerida are Ceractinomorpha with a skeleton which is always composed of a combination of spicule and spongin fibre. The megascleres are monactine or diactine with many curious structural variations. Spiny spicules are common. Both fibre and spicule skeletons can be complex and show great regional differentiation. Microsclere types are varied, basically chelate (Pls. 6*c*, *d*, 7*a*), sigmoid (Fig. 3.4*b*, p. 93), or toxiform (Fig. 3.4*k*, p. 93; Pl. 7*d*), again with many detailed structural variants. The larvae are parenchymellae with incomplete ciliation, the posterior pole is always bare (Pl. 12*b*), and anterior and posterior poles may show differential pigmentation.

Family **Mycalidae** Lundbeck

Poecilosclerida with a diffuse plumoreticulate spicule and fibre skeleton in which the megascleres are styles or subtylostyles. The microscleres always include anisochelae (Pl. 7*a*) to which may be added sigmas, toxas, raphides and isochelae of many types.

Mycale (Fig. 5.18), *Paraesperella.*

Family **Hamacanthidae** Gray

Poecilosclerida with an irregular megasclere skeleton of smooth monactinal spicules, accompanied by peculiar, sharp-toothed, cheloid microscleres, termed diancistras.

Hamancantha.

Family **Cladorhizidae** De Laubenfels

A group of small, deep-water poecilosclerids with a heavily siliceous skeleton made up of styles organized into an axis from which radial tracts diverge. The shape of the body is often complex, and microscleres, when present, are chelae and sigmas.

Chondrocladia, Cladorhiza.

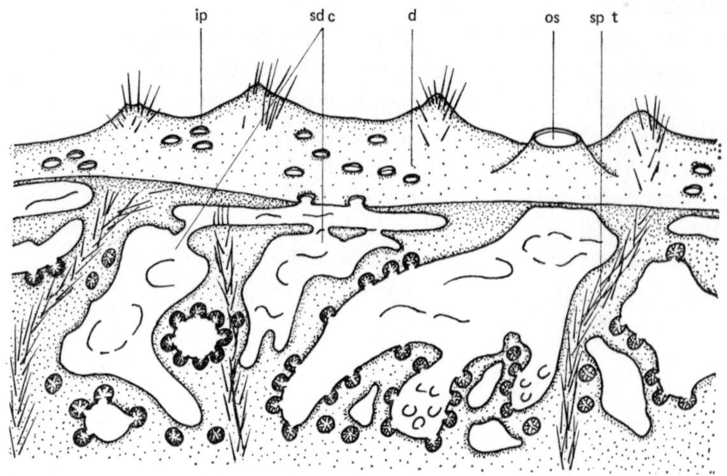

Fig. 5.18 Diagrammatic section of *Mycale* (Poecilosclerida) cut at right angles to the surface.
os, oscule; spt, plumose spicule tracts; sdc, subdermal cavities; d, dermal membrane; ip, inhalant pore.
(Redrawn from Hentschel 1923.)

Family **Biemnidae** Hentschel
Poecilosclerida with skeleton of smooth styles organized in plumoreticulate fashion, but occasionally showing some axial condensation in stalked, lamellate forms. The microscleres are abundant and diverse, including sigmas, microxeas and toxas, sometimes with curious siliceous spheres and 'commas'.
Biemna, Desmacella.

Family **Desmacidonidae** Gray
Poecilosclerida with reticulate or plumoreticulate skeleton and diactinal megascleres which are of uniform type throughout the sponge. Microscleres can be both sigmoid and chelate, the latter often of very curious form.
Desmacidon, Guitarra.

Family **Coelosphaeridae** Hentschel
Poecilosclerida generally of spherical shape with inhalant and exhalant openings reduced to specialized tubular fistules and the remainder of the surface a dense layer of tangentially disposed diactinal megascleres. The choanosome of these sponges is a very delicate network of plumoreticulate spicule fibres which always collapses as soon as the sponge leaves the water. Thus it has frequently been

referred to as 'pulpy' and structureless. Microscleres are chelae, toxas, sigmas and microxeas.

Inflatella, Coelosphaera.

Family **Crellidae** Hentschel

Poecilosclerida with a principal skeleton of diactinal megaslceres arranged in plumose or plumoreticulate fibrous tracts, often echinated by acanthostyles. The dermal skeleton is a dense layer of acanthostyles. Encrusting forms have a layer of erect acanthostyles aligned along the attachment surface. Microscleres are arcuate isochelae (Pl. 6*d*) and sigmas.

Crella, Pytheas.

Family **Myxillidae** Topsent

Sponges with a regular, reticulate skeleton of monactinal megascleres which are styles and acanthostyles to which echinating acanthostyles may be added (Fig. 1.8*e*, p. 46). The dermal spicules are diactinal and the microscleres of various types, but most frequently arcuate isochelae and sigmas.

Myxilla, Ectomyxilla (Pl. 6*d*), *Lissodendoryx* (Fig. 5.19).

Family **Tedaniidae** Ridley and Dendy

Poecilosclerida with an endosomal skeleton of styles which are organized into plumoreticulate tracts. The ectosomal megascleres are diacts, usually tylotes. Microscleres are onychaetes, which are extremely thin, long, oxeote microscleres with a roughened surface.

Tedania.

Family **Hymedesmiidae** Topsent

Poecilosclerida of permanently encrusting growth form with both principal and accessory endosomal spicules ancanthostyles, which are oriented vertically to the attachment surface in an arrangement termed 'hymedesmoid' (Fig. 1.8*c*, p. 46). Ectosomal spicules are either diactinal or monactinal, oriented vertically, or strewn without organization throughout the thin body. They are usually more slender than the principal spicules. Microscleres include arcuate or unguiferate isochelae and sigmas, in addition there are many peculiar forms which serve to characterize the genera. The surface of the sponge very frequently has specialized oscular and pore areas.

Hymedesmia, Stylopus.

Family **Anchinoidae** Topsent

Poecilosclerida with principal endosomal skeleton composed of plumose to plumoreticulate columns of diactinal spicules echinated by smooth or spiny styles. Ectosomal spicules are diactinal, often of the same type as those of the endosome. Microscleres are isochelae and sigmas. There is no dermal spicule crust and spongin fibre is not a major component of the skeleton. These negative

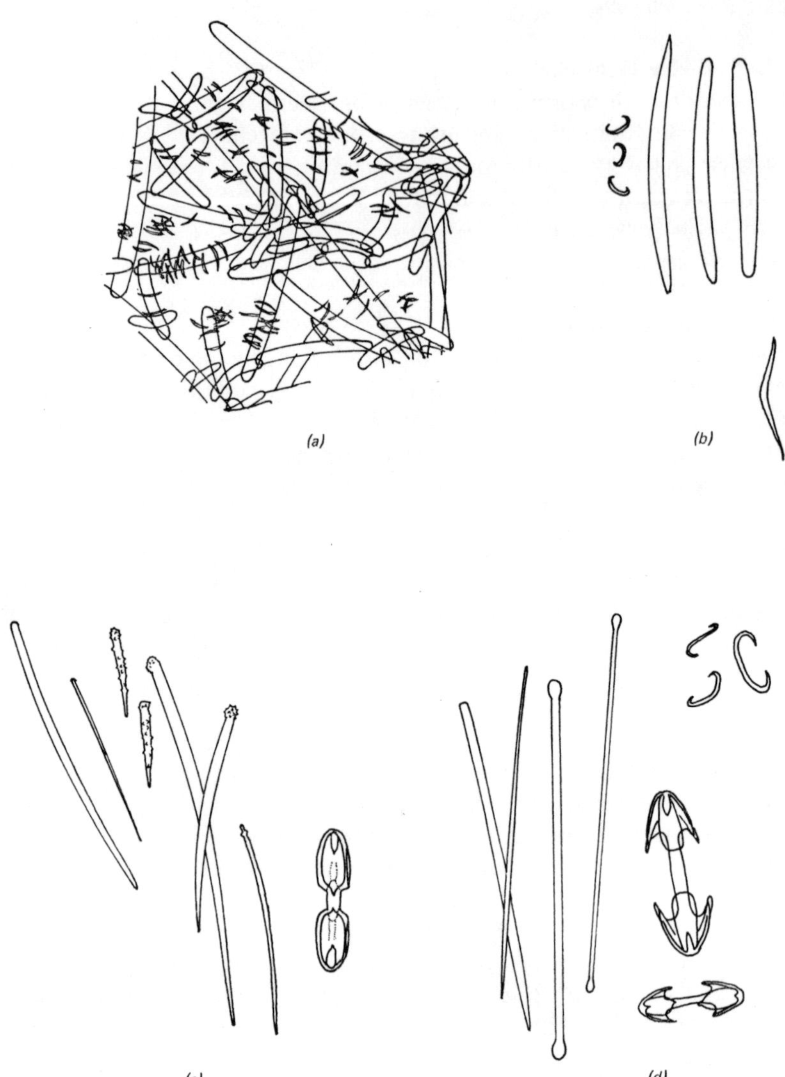

Fig. 5.19 Spiculation of some typical members of the Ceractinomorpha. (*a*) *Strongylophora durissima* (Haplosclerida, Adociidae); (*b*) *Orina regius* (Haplosclerida, Adociidae); (*c*) *Microciona prolifera* (Poecilosclerida, Clathriidae); (*d*) *Lissodendoryx isodictyalis* (Poecilosclerida, Myxillidae). (*a*, redrawn from Dendy, 1905; *c*, *d*, redrawn from Hartman 1958*a*.)

characters set this family apart from the Crellidae with which it may otherwise be confused.

Anchinoe, Hamigera.

Family **Clathriidae** Hentschel

Poecilosclerida with a skeleton in which styles or acanthostyles are the principal spicules. These are organized into spicule tracts which include variable quantities of spongin and which support echinating acanthostyles. Accessory ectosomal spicules are also styles, usually finer than the endosomal types (Fig. 1.8*f*, p. 46). There are no diactinal megascleres. Microscleres are palmate isochelae and toxas.

Clathria, Microciona (Fig. 5.19*c*), *Axociella* (Pl. 7*d*).

Order **Haplosclerida** Topsent

Demospongiae with a reticulate skeleton in which the pattern is isodictyal with rectangular or triangular meshes, which can be unispicular, multispicular or constructed entirely of fibre with spicules lacking.

Principal spicules are normally oxeas or strongyles of relatively uniform length within a species. Microscleres, when present, are sigmas and toxas.

There is no regional size differentiation in the mineral skeleton, but a distinct tangential dermal spicule or fibre skeleton is developed in some families.

Larvae are incompletely ciliated; the posterior pole is bare, often pigmented and is fringed by a ring of longer cilia (Pl. 12*d*).

Family **Haliclonidae** De Laubenfels

Haplosclerida with spicules of uniformly small size organized into an isodictyal skeleton which is always consolidated by spongin, either at the intersection of spicule meshes, or in the form of fibres (Fig. 1.8*a*, *b*, p. 46). Spongin can become much more prominent than spicule in the skeleton. Microscleres are typically absent and there is never any specialized tangential dermal skeleton.

Haliclona.

Family **Adociidae**

Haplosclerida with diactinal spicules arranged in isodictyal fashion, frequently in polyspicular tracts. Oxeas and strongyles are common and occur in a range of sizes which are not, however, confined in size categories to specific regions. Microscleres, when present, are sigmas and toxas. All members of the family have a tangential dermal spicule skeleton and many emphasize the siliceous spicules rather than the fibres, and are thus very crisp and brittle.

Adocia, Strongylophora (Fig. 5.19*a*), *Orina* (Fig. 5.19*b*), *Gellius* (Pl. 8*c*).

Family **Callyspongiidae** De Laubenfels

Haplosclerida with oxeote or strongylote spicules largely enclosed inside spongin fibres. In some cases no spicules are present. All species have a massive spongin

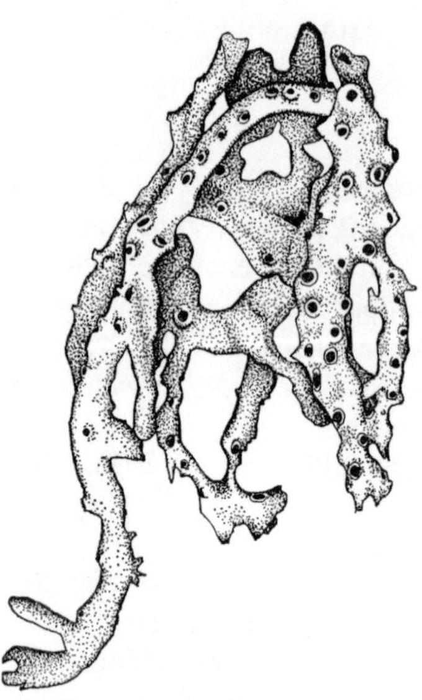

Fig. 5.20 Callyspongia. This illustrates a typical growth form for many Haplosclerida, particularly those with fibrous skeletons belonging to the Haliclondiae and Callyspongiidae. They are often repent and branching with slightly elevated oscules.

skeleton which may incorporate detritus instead of spicule. A specialized tangential dermal reticulation of spongin fibres is present and this incorporates primary, secondary and sometimes even tertiary meshes. Microscleres, when present, are toxas.

Callyspongia (Fig. 5.20), *Dactylia* (Pl. 5*b*).

Family **Spongillidae** Gray

Fresh-water sponges, with a soft cavernous structure which is most noticeable just below the surface. Spicules are smooth or spined oxeas, or strongyles, organized into bundles or tracts and bound by spongin. Production of gemmules is characteristic, but not diagnostic. However, the complex structure of the gemmule coat with its so called pneumatic layer of loosely packed spongin is unique to fresh-water sponges (Fig. 3.1, p. 86). Particular, complex microscleres are often associated with the gemmule coat.

Spongia, Ephydatia.

Family **Potamolepidae** Brien

Fresh-water sponges with a rigid reticulate skeleton formed of bundles of curved strongyles, to which ectosomal and dispersed endosomal oxeas may be added.

Asexual reproductive bodies are defined as statoblasts, not gemmules, as they lack the complex pneumatic layer of the typical gemmule and are covered by endosomal strongyles and microscleres. They are more similar in structure to gemmules of some marine Haplosclerida than to those of other fresh-water sponges.

Potamolepis.

Order **Dictyoceratida** Minchin

Ceractinomorpha in which the spicule skeleton is lacking and is replaced by a spongin fibre skeleton often of great complexity. This is always constructed upon an anastomosing pattern, involving recognizable primary and secondary fibres (Fig. 5.21).

Larvae are often very large, up to 5 mm, with ciliation similar to that of the Haplosclerida, although the posterior cilia can be much longer (Fig. 5.22).

Family **Spongiidae** Gray

Dictyoceratida with small flagellated chambers around 20–40 μm in diameter and fibres which are homogeneous in cross-section, not having a diffuse central pith.

Spongia, Phyllospongia.

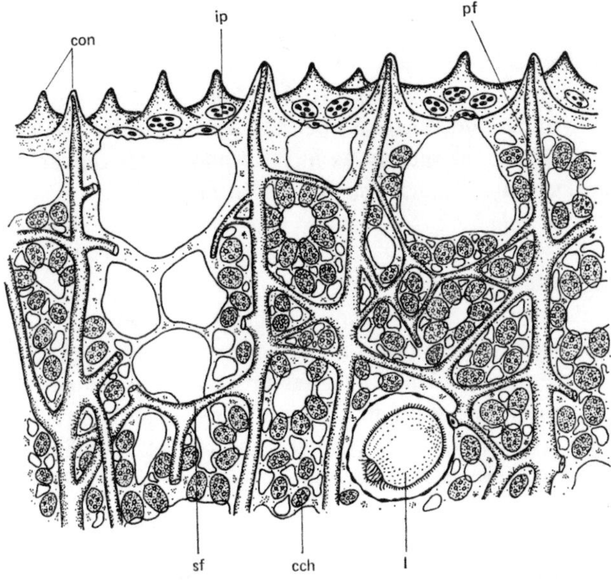

Fig. 5.21 Diagrammatic section at right angles to the surface of a dictyoceratid sponge. Primary ascending spongin fibres and connecting secondary fibres make up the network. The elevation of the dermal membrane into surface conules is shown.
pf, primary fibre; sf, secondary fibre; l, larva; cch, choanocyte chamber; ip, incurrent pores; con, conules. (Redrawn from Dendy 1905.)

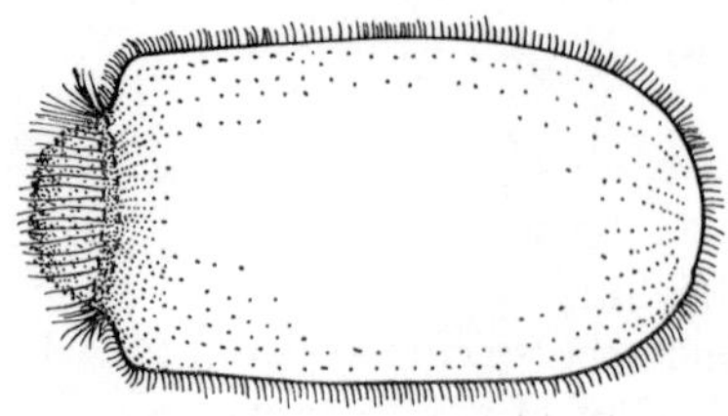

Fig. 5.22 Parenchymella larva of *Spongia reticulata* (Dictyoceratida).

Family **Thorectidae** fam. nov.
Dictyoceratida with small flagellated chambers 20–60 μm in diameter and fibres often of very large size which have a distinct stratified structure in cross-section. The spongin skeleton forms a very regular rectangular mesh with wide spaces between fibres.

Thorecta, Halispongia.

Family **Dysideidae** Gray
Dictyoceratida in which the flagellated chambers are very large, over 50 μm, and often oval with a wide exhalant opening, a shape termed 'eurypylous'. The fibres are homogeneous in section as in the Spongiidae, but do not form the massive tough skeletons characteristic of the latter group.

Dysidea, Euryspongia.

Order **Dendroceratida** Minchin
Ceractinomorpha which lack siliceous spicules and which have a skeleton composed entirely of spongin fibres arranged on a dendritic pattern (Fig. 5.23), only rarely anastomosing. This can sometimes be supplemented by spongin spicules. In one family the spongin skeleton is lost completely.

Larvae of the Aplysillidae are similar to those of the Haplosclerida and Dictyoceratida, but those of *Halisarca* lack long posterior cilia.

Family **Aplysillidae** Vosmaer
Dendroceratida in which the spongin skeleton is present.

Aplysilla (Pl. 12*c*), *Darwinella.*

Family **Halisarcidae** Vosmaer
Dendroceratida in which the spongin skeleton is absent.

Halisarca (Fig. 5.24), *Bajalus.*

Fig. 5.23 Fibre skeleton of *Dendrilla rosea* (Dendroceratida).
The dendritic pattern is typical of the order, but most members have less complex body form, with an encrusting habit being common. (Redrawn from Lendenfeld 1889.)

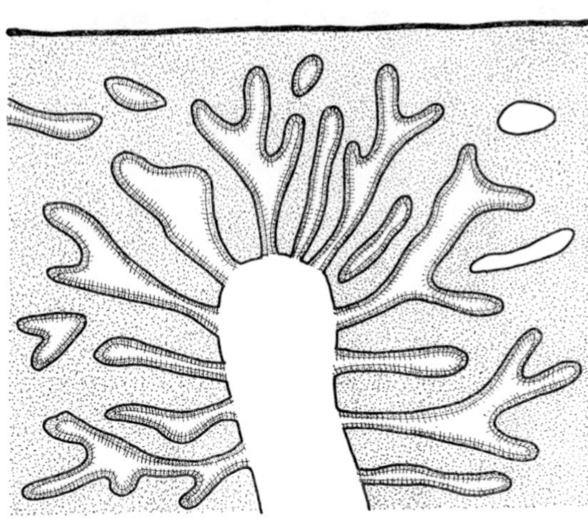

Fig. 5.24 Diagrammatic section of *Halisarca* cut at right angles to the surface.
The choanocyte chambers are long, branching sacs.
(Redrawn from Lenenfeld 1889.)

Order **Verongida** ord. nov.
Ceractinomorpha which lack a mineral skeleton, and have a reduced spongin fibre skeleton which supports a heavily collagenous matrix. Fibres have a pith and in some cases the pith alone is represented, and the outer bark lost. Choanocyte chambers are small and spherical, 20–30 μm in diameter. Electron microscope studies have revealed the verongiids to have complex histology with many cell types which are quite divergent from those of other Ceractinomorpha.

In so far as development is known, the Verongida are oviparous. This is completely contrary to their inclusion in the Ceractinomorpha, but until a complete account of their development becomes available, they are retained in this subclass. All histological and biochemical evidence supports their wide separation from other aspiculous Demospongiae, probably as a separate subclass.

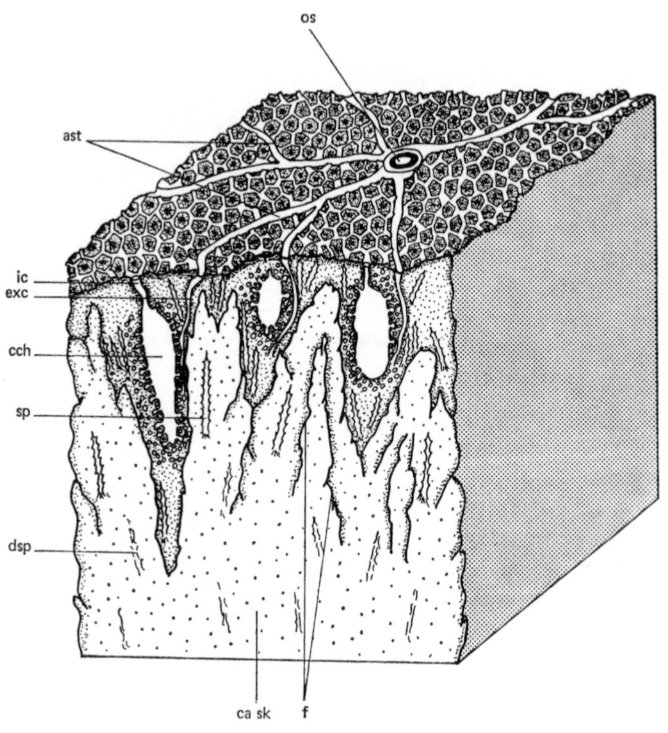

Fig. 5.25 Diagram to show the typical morphology of a sclerosponge belonging to the order Ceratoporellida. One face and the upper surface are represented. ic, incurrent canal; cch, choanocyte chambers; sp, siliceous spicule; dsp, eroding spicule; ca sk, calcareous skeleton; f, spongin fibres; exc, excurrent canal; os, osculum; ast, astrorhiza. (Based upon Hutchinson, 1970.)

Family **Verongiidae** De Laubenfels
Diagnosis as for the order.
Verongia, Psammaplysilla (Pl. 4*a*), *Ianthella.*

5.6 Class Sclerospongiae Hartman and Goreau

Sponges which secrete a compound skeleton of siliceous spicules, spongin fibres and calcium carbonate, the latter deposited as a basal mass in which the siliceous spicules may or may not become entrapped. Astrorhizal patterns are evident on the calcareous surface (Pl. 11*b*; Fig. 5.25).

The living sponge tissue is entirely comparable to that of the Demospongiae, as indeed are some of the spicule types. However, the tissue is divided into units, each of which extends down into the upper layer of the basal calcareous skeleton. The remainder of the calcareous skeleton is cut off from the living tissue by either tabular structures or a solid calcareous deposit. The microstructure of the aragonite is trabecular.

Order **Ceratoporellida** Hartman and Goreau
Recent Sclerospongiae with a compound skeleton composed of a basal mass of aragonite, spongin fibres and siliceous spicules (Fig. 5.26) that become entrapped in the calcareous skeleton. Living tissue forms a veneer on the surface of the basal calcareous mass and extends down into calicular skeletal units. Calicles multiply by longitudinal fission and are filled in solidly with secondary deposits of calcium carbonate.

Ceratoporella (Fig. 5.25), *Astrosclera* (Pl. 11*a*, *b*), *Merlia* (Fig. 5.27).

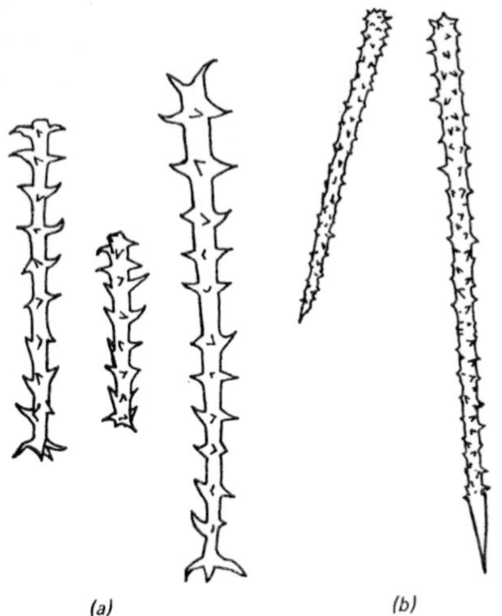

Fig. 5.26 Siliceous spicules of Sclerospongiae.
(*a*) *Goreauiella auriculata* (Ceratoporellida); (*b*) *Stromatospongia vermicola* (Ceratoporellida). (Redrawn from Hartman 1969.)

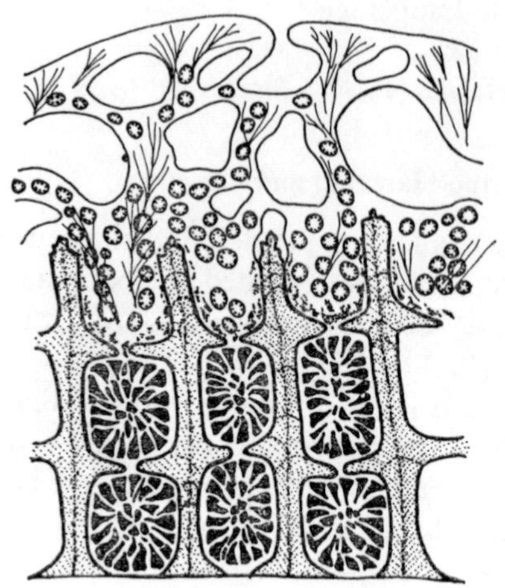

Fig. 5.27 Vertical section of *Merlia normani* showing the basal calcareous matrix in which the individual compartments are filled by secondary deposition. The superficial soft tissue contains the choanocyte chambers and is supported by tracts of siliceous spicules. (Redrawn from Kirkpatrick 1911.)

Order **Tabulospongida** Hartman and Goreau

Sclerospongiae with a calcitic basal skeleton which has a lamellar microstructure. The walls of the calicles may be provided with spines, trabeculae can be secondarily thickened and of irregular arrangement. Asexual reproduction is by intramural budding. The living tissue of recent forms secretes siliceous spicules which are tylostyles and spirasters similar to those of the Clionidae and Spirastrellidae.

Acanthochaetetes.

6 Sponge ecology

6.1 Introduction

Sponges have colonized all aquatic habitats. Some groups, such as the Calcarea, are restricted by physical factors that limit skeleton secretion to a relatively shallow zone from the intertidal to 100 m; others are preferentially inhabitants of deep water, for example the Hexactinellida. Demospongiae can be found in any situation, from upper intertidal to hadal depths (down to 8600 m), in fresh and brackish water, in caves and in full illumination. While Calcarea are restricted to firm substrates, and many Hexactinellids colonize soft surfaces, Demospongiae can utilize rock, unstable shell, sand and mud substrates, and in some cases they excavate into calcareous material. The ecological dominance of Demospongiae is simply a reflection of their diversity in form, structure, reproductive capabilities and physiological adaptation.

It is always a problem to decide what constitutes ecology; where, if at all, to draw the line between ecology and physiology. In the case of sponges, where so many crucial details of physiology and reproductive biology remain to be elucidated, it is very difficult to take an 'ecological' viewpoint. However, some simple facts which apply to most sponges can be taken as a base-line for ecological interpretation. Despite differences in range and habitat between different taxa these features, which are pertinent to ecology, are common to all groups. All sponges are microphagous filter-feeders and most are sedentary, if not immobile, in the adult stage. In the course of reproduction most sponges have a free stage, a larva, which has a brief independent existence. Factors which influence the distribution and survival of both life history stages must be considered when studying sponge ecology.

6.2 Ecology of sponge larvae

The presence of a free larva at some stage in the life history is typical for Porifera and where exceptions are known, as for example in the Spirophorida, the suppression of the larval stage relates clearly to the requirements of particular habitats. While our knowledge of the morphology and morphogenesis of sponge larvae is now sufficiently advanced to be useful in classification and to contribute to interpretation of phyletic relationships in most groups, relatively little information is available on the behaviour of the larva during its free existence. All sponge larvae are ciliated and small, ranging from 50 μm in length in the amphiblastula of *Sycon* to 5 mm in the parenchymella of *Stelospongia*. They may have regions of longer cilia, or areas which lack cilia, the latter areas frequently coincide with a differential pigmentation of posterior or anterior poles. As planktonic organisms, sponge larvae are remarkably simple and lack any obvious structure which might allow them to exert behavioural preferences in response to environmental conditions. Cellular differentiation can be well advanced in larvae, but the cells are generally of the adult type. The only specialized larval cells are those of the ciliated epithelium and the '*cellules en croix*' of some Calcaronea.

Despite their simplicity, sponge larvae do show a range of responses to light and gravity, they have a variety of types of movement, they differ greatly in longevity and in their reaction to contact with settlement surfaces. These behavioural attributes are of vital importance to the study of the ecology of adult sponges, since they influence dispersal and to some degree enable habitat selection.

6.2.1 Emission of larvae and gametes

One obvious factor in larval ecology is the nature of larval release and, in oviparous species, the nature of gamete release. It is clear that some sponges are quite asynchronous within individuals and within populations in this respect, while other species appear to synchronize reproductive events so that larvae, or eggs and sperm, are released in huge numbers at one time (Pl. 4*c*). There is no clear pattern for any taxon or habitat at this time. Obviously epidemic release maximizes the possibility of successful fertilization in synchronized oviparous species. It is difficult to see any advantage in epidemic release of incubated larvae throughout a population. Given the capricious conditions of intertidal and shallow subtidal habitats, larvae which are reasonably short lived, and which do not swim actively, may achieve a higher percentage survival from a phased pattern of release. There is no convincing evidence that, in a natural habitat, water movement or its arrest stimulates release of reproductive products.

6.2.2 Larval movement

Three distinct types of movement have been observed in sponge larvae:

(i) Directional swimming with constant rotation. This is performed by larvae such as the parenchymella of *Haliclona*, which, aided by a posterior ring of long cilia, is capable of fast directional movement at a rate of 1 cm^{-1}.

(ii) Swimming on a spiral path while rotating constantly. This type of movement has been detected in all species in which larvae swim actively and has been termed 'corkscrew'.

(iii) Larvae of some species, for example *Halichondria moorei*, never swim but creep over a surface while rotating about their long axis.

Such behaviour during the free stages has obvious significance in any attempt to interpret dispersal patterns of adult sponges. In most species the larvae pass briefly through a creeping stage just prior to settlement.

6.2.3 Duration of free life and settlement behaviour

The shortest period of free life is found in the larvae of the Clathriidae (*Microciona, Ophlitaspongia*) where active swimming for 2–3 hours is followed by crawling for a similar period. Attachment is by the anterior pole, and cellular reorganization and flattening is rapid: the larva is a flat disc within 1 hour, and a canal system with a single osculum is functional within 2 days.

Many larvae of the order Halichondria have a crawling phase of up to 60 hours before settlement and after that have an organized canal system within 72 hours. The small, undifferentiated creeping blastula of *Polymastia* has a free life of 18–20 days, and then an extended settlement period which may last 7 days. The period of reorganization required to produce a functional sponge is unknown.

Sponge larvae settle and metamorphose by attaching at or near the anterior pole. The site of attachment is defined before settlement by the appearance of an area of secretory cells either at the anterior pole (Pl. 12*c*), or just posterior to it, within the ciliated layer.

It is difficult to be dogmatic when dealing with material as plastic as sponge larvae. They will certainly treat unsuitable surfaces such as an air/water interface as a settlement substrate and begin to metamorphose there and can also adapt to other influences by abnormal behaviour. Reports of settlement of larvae on their side or by the posterior pole can be considered in this light.

6.2.4 Reactions of larvae to physical stimuli

If one ignores the gross environmental factors such as temperature and salinity which can set absolute limits for species distribution, then three physical attributes appear to influence the swimming and settlement of sponge larvae: light, gravity and water turbulence. In all cases a variety of responses of larvae to these factors has been reported. The response of some larvae to light gradients is extremely marked. It is a common pattern to find that a larva which is positively phototactic throughout its period of active swimming will become indifferent to light, or even negatively phototactic, during the pre-settlement creeping phase. An example given by Bergquist *et al.* (1970) is *Haliclona* sp. The same authors recorded strong negative phototaxis in *Mycale macilenta* larvae during the swimming period, these larvae show a spectacular response to light and shade; as they approach a boundary they turn through 180° and swim back into the shade. They will do this repeatedly. It is not known if this behaviour is reversed before settlement, but observations on the habitat of the adult sponge, a cryptic species, suggest that it is not. Some larvae, for example those of *Microciona coccinea* and *Halichondria moorei* are indifferent to light gradients.

Gravitational responses have been reported by Warburton (1966) for *Microciona prolifera* and *Cliona celata* larvae. During the first 24 hours of free life they display negative geotaxis, but between 20 and 30 hours after liberation this is reversed and the larvae seek the substrate and settle. Continuous turbulence will upset this pattern, suppress swimming, and keep the larvae on the substrate throughout the free period.

6.2.5 Substrate selection

The larvae of many sessile marine invertebrates are unselective as to the nature of the surface on which they settle, provided, in the case of artificial surfaces, that a prior bacterial or algal film has been established. This is true for sponge larvae, which prefer a pre-coated surface. For some, such as *Ophilitaspongia seriata*, this is an absolute requirement.

Most species tested prefer natural rock surfaces such as basalt, bivalve shells or coralline algae to artificial surfaces such as glass or plastic. However, *Cliona celata* larvae have been recorded as showing no preference for calcite over glass. This is surprising since larvae of the boring clionids must settle on a calcareous surface to survive. An earlier study (Hartman 1958*a*) in which larvae of the same species showed a clear preference for settlement on the valves of oysters is more likely to be correct.

In general, however, sponge larvae appear to be able to settle on a variety of surfaces. In those cases where very specific associations between sponges and settlement surfaces are known, for example *Microciona coccinea* investing the shell of the gastropod *Herpetopoma bella* and *Dactylia palmata* and *Iophon minor* investing *Chlamys zealandiae*, the sponge always has a wide distribution and is not limited to the specific surface, although it may be the only sponge that will grow on that host.

It is probable that the orientation of a particular settlement surface is more significant than its composition. For example, the non-swimming larvae of *Halichondria moorei* will always creep up the sides of a tank and settle on any surface available that is located at, or just below, the air/water interface. This is remarkable when the location of adult colonies of this intertidal species is considered. They occur fringing boulders in standing water or fringing midtidal pools; larvae which on release seek the air/water boundary and then settle quickly, regardless of light gradients or substrate, would ensure such distribution.

6.2.6 Larval mortality

Because it is difficult even under experimental conditions to monitor the output of larvae from an individual, and to assess the proportion of those larvae which first settle and then survive, there is little precise information on larval mortality. Heavy predation of free larvae by spionid polychaetes is known, and opisthobranchs and pomacentrid fish predate settled larvae. High percentage larval settlement, in relation to observed larval release, has been observed for *Halichondria moorei* and *Crella incrustans*.

One life table for a sponge has been assembled by Hartman (1958*a*) for *Haliclona loosanoffi*.

Table 6.1 *Life table for newly settled colonies of* Haliclona loosanoffi.

x Age in months	d_x No. dying in interval out of 1000 settled larvae	1_x No. surviving in beginning of age interval out of 1000 settled larvae	$1000q_x$ Mortality rate per 1000 alive at beginning of age interval
0–1	564	1000	564
1–2	154	436	353
2–3	0	282	0

The data indicate heavy mortality in the early stages and long survival of a few individuals which live to advanced age.

6.2.7 Fusion of sponge larvae

Several workers have reported the fusion of sponge larvae at or immediately after settlement. The cell mass which then goes on to produce a single individual can be augmented in this way, and Fry (1971) has suggested that the greater cell volume actually enhances the possibility of survival. In his experiments on *Ophlitaspongia seriata*, fusion masses composed of two to four larvae survived in higher proportion than did sponges which developed from a single larva. Fusing larvae in this case were derived from a single parent. Whether *Ophlitaspongia* larvae of separate parentage, or from different sponge populations, will fuse was not determined. Despite these laboratory-based observations that fusion can occur, there is no evidence to suggest that successful settlement and growth of sponge larvae is dependent on fusion. There are many records of single larvae producing functional sponges.

There is no information available on whether the future reproductive potential of individuals derived from multiple larvae is greater than those derived from single larvae. To infer that larval fusion had any selective advantage one would need to confirm that there was a reproductive advantage in terms of quantity of larvae released or an advantage in terms of increased heterozygosity among the gametes produced. The latter is unlikely, since fusion is known to be successful only between larvae of the same clone.

There is good evidence (Van der Vyver 1970) that larvae of *Crambe crambe* will not fuse if they derive from separate parents, but in 75 % of cases will fuse on contact if they derive from the same parent at the same time. That is to say, fusion requires that they are in similar growth phases at the time of contact.

6.2.8 Relationships between larval behaviour and adult sponge habitat

From the preceding account it is clear that sponge larvae can respond in a variety of ways to most environmental stimuli. Consequently, it has always been assumed that the observed habitat specialization of adult individuals results from selective mortality following unselective settlement. While this appears true when only selection of substrate by its chemical nature is considered, it is not true when one considers the highly species-specific behaviour demonstrated for larvae such as *Halichondria moorei* and *Haliclona* sp. In these cases the combination of swimming or crawling behaviour, duration of free life, and phototactic response, combined with unspecific requirements as to settlement surface, works to minimize the dispersion of the larva into unsuitable habitats. Such patterns are only known for intertidal sponges, but are likely to be common and of great significance to the interpretation of adult distribution patterns.

6.3 Ecology of the sessile stages

6.3.1 The influence of physico-chemical factors

Many physical and chemical factors of the marine environment affect the distribution, metabolism and reproductive efficiency of sponges. However, there is little to be found in the literature on these subjects to support any general statements. To be intelligible, any treatment would have to be detailed. Sara and Vacelet (1973) in their contribution to the *Traité de Zoologie* have included an excellent coverage of the literature of physico-chemical factors and their effects on marine sponges and, therefore, since nothing new needs to be added, the reader is referred to their article.

6.3.2 Sponges in their habitats

An interesting and active area of ecological investigation at present concentrates on sponges in their habitats, emphasizing the biological attributes sponges bring to bear in obtaining and maintaining their specific positions on the shore or on submerged surfaces. Of interest are the special problems faced by sedentary benthic invertebrates in their interactions with other organisms, and the sponge solutions to these problems.

Depending upon their precise habitat the problems of life for sedentary invertebrates differ substantially. All must find attachment space, however, and this must be sufficient to allow development of adequate numbers of individuals, or large enough body size, to sustain reproduction in the population at a viable level. Having obtained suitable space, this resource, which in marine environments is generally more frequently limiting than food, must then be protected against competitive inroads by other species. At the same time the sedentary organism is vulnerable to predation by mobile organisms and presents a surface on which larvae either of predators or of spatial competitors may settle and overgrow. Sessile organisms also are to a varying degree dependent on a free larval stage for dispersal and normally such stages face great attrition of numbers in the period before settlement.

For obvious reasons most experimentation on sessile invertebrates, and how they interact over a period of time with other organisms, has been focussed on intertidal communities, where caging and disruption experiments with adequate control areas can be set up easily. It is now clear that on such intertidal hard surfaces, if predation and disturbance of the environment are prevented, then a single competitive dominant species will take over and occupy the available space. Therefore, to maintain high species diversity in such locations a predator, or physical disturbance, such as storm effects or ice scour, must be invoked. In intertidal locations sponges are not physiognomic organisms; they occur frequently, in pools, in shade, under boulders, in crevices, or between barnacles, but do not cover large areas. In this region sponges need not seek special microhabitats. They are not the major space occupiers. The special situations of complex sponge communities on wharf pilings and mangrove roots can be interpreted as examples of subtidal communities elevated in level by shaded conditions. Therefore the main question to be answered is: do the generalizations for the intertidal types hold for the submerged hard substrates on which, particularly in low light, sponges are clear spatial dominants? If this were so, we would expect to see in such communities obvious predators controlling growth of the sponges and to observe that competitive dominants emerged with time.

However, although little has been published which deals specifically with this question, the observation of this author after many years of field experience in shallow subtidal sponge habitats is that selective sponge predators are rarely obvious and the influence of unselective predation is minimal. Physical disturbance is unusual, diversity is high, patchiness extreme and competitive dominance rare. These observations are supported by those of Sara (1970) and Jackson and Buss (1975). Also, over periods of five years, particular complex environments on shaded vertical faces have been monitored and show no change in species diversity and little alteration of species boundaries. The submerged hard surface communities, where sponges often constitute a continuous cover, give the impression of longevity and great stability.

There are some exceptions. One is to be found on boulder surfaces, where Rützler (1965), analysing the number of species in a given area on boulders of varying weight, found that species diversity decreased with increasing rock weight between 3 kg and 30 kg. From 30 kg to stable rock it remained constant. This could indicate that interspecific competition for space commenced when rock size allowed some stability in the face of wave disturbance. This point was reached at 3 kg, and from this boulder weight upward the faster-growing species were interpreted as being able to exclude their slower growing neighbours. However, a boulder community, where even large boulders are known to move, offers at best a precarious habitat and has many of the characteristics of an intertidal habitat where disturbance in the form of overturned boulders is common. The fact that the large, stable surfaces do not show this trend toward reduction in diversity suggests that disturbance, not interspecific competition, is the shaping force in this habitat, and that it is promoting the reverse of what happens in the intertidal, that is disturbance here is reducing diversity.

Another exception is to be found in the work of Dayton, Robilliard, Paine and Dayton (1974), who worked on an Antarctic, sponge-dominated community, between 30 and 60 m depth, below the zone of disturbance by anchor ice. This is a community in which the sessile species, which are mainly sponges, share attachment space as the most significant and potentially limiting resource. There are also some very obvious echinoderm and opisthobranch predators. Thus here it proved possible to isolate the effects of interspecific competition and predation in a notably disturbance-free, benthic environment.

Several points emerged from this study. Growth rates of all except one rather rare sponge, *Mycale acerata*, were slow; in most cases so slow that a year showed no change. The large Hexactinellida were estimated to be several hundred years old! Using superior growth rate as a measure of competitive dominance, Dayton concluded that *Mycale* should, in reality, occupy more of the habitat than it did. It is known to produce free-swimming larvae in large numbers and can thus disperse onto the available free attachment surface, which in this case was 40 % of the bottom area. Dayton set up caging experiments and demonstrated that *Mycale* was prevented from monopolizing the available space by the selective predation of two asteroids, species of *Perknaster* and *Acodontaster*; inside cages which excluded the starfish there was runaway growth of *Mycale*.

The most abundant sponges in this community were three hexactinellids and *Tetilla leptoderma*. Caging experiments showed that three individuals of *Acodontaster* could consume 55 % of a very large specimen of *Rossella nuda* in one year. These rossellid sponges are large enough for a person to sit inside them! Thus large size and heavy spicule content do not eliminate the threat of predation. Presumably, the high standing crop of these four sponges in this Antarctic environment is permitted by some source of selective mortality in the predator species. This is possibly provided by the abundant, detritus-feeding asteroid *Odontaster validus*, which could consume many larvae of the other sponge-feeding echinoderms.

Another common sponge was *Cinachyra antarctica*. It was not eaten at all by starfish, although in structure, abundance and percentage organic matter available to a consumer, it is closely comparable to *Tetilla leptoderma*, which was the most preferred sponge for the two starfish predators. No suggestion was advanced by Dayton to explain this, but subsequently we have demonstrated that several species of *Cinachyra* produce toxic terpenoid compounds. It is thus probable that *Cinachyra* utilizes a chemical defence against predation.

Dayton *et al.* (1974) have demonstrated beautifully the importance of predation in the dynamics of a long-established, undisturbed benthic environment. The same theory cannot, however, be extrapolated to include habitats where space is absolutely limiting, as in the sponge-dominated habitat in caves, under boulders, and on vertical faces where free space is as low as 1 % and predators are not obvious. Dayton did not in fact show that spatial competition was occurring in the Antarctic community, and many features of larval production and behaviour in individual species need to be elucidated before one could determine this. Also, it is possible that predation in this instance was keeping density so low that spatial competition was unimportant, regardless of any restrictions on larval settlement that might exist.

Implicit in both of these studies is the idea, derived from terrestrial and intertidal ecological models, that competitive relationships between species will be established as linear hierarchies.

In order to interpret in dynamic terms how spatial patterns of great complexity are produced and maintained in other sponge-dominated habitats, it is necessary to look closely at the biological and biochemical attributes of the organisms which make up the communities. How do sponges co-habit, over long periods at high levels of diversity and low levels of both predation and disturbance? What are the particular features of sponges which are relevant to their interactions with other sponges and sessile invertebrates in space-limited habitats?

The extreme flexibility in growth form that many sponges possess poses problems in description, but, coupled with the non-regional organization of the sponge body, permits many adjustments as growth proceeds. They are constrained by few rigid growth requirements. Within the phylum, as we have seen, there is a wide range of both sexual and asexual reproductive patterns. There is also considerable variability in reproductive period. These factors confer great flexibility on sponges with regard to the time an individual can remain on, and extend over, a surface or colonize a substrate by new settlement.

Growth rates, particularly the time taken from initial settlement to initial reproduction can vary widely. Some sponges, for example *Halisarca*, have been

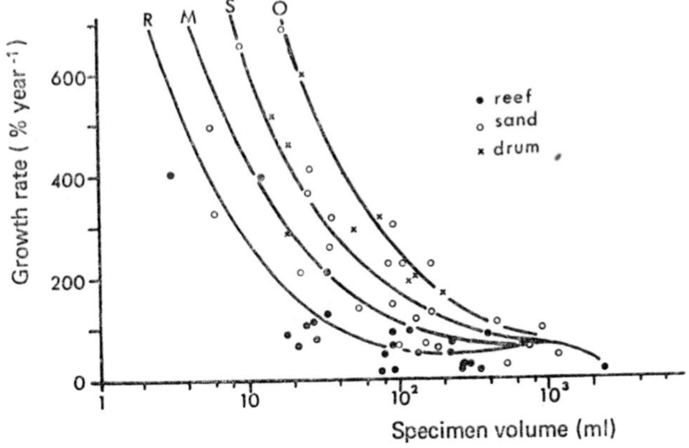

Fig. 6.1 The growth rate of *Mycale*.
Measured rates of annual growth are indicated for specimens in three habitats: solid reef (●), sand (○) and on a metal drum (x). Smoothed growth versus size curves are drawn for reef (R) and sand populations (S), for the mean of the entire population (M) and for optimum conditions (O). (After Reiswig 1973.)

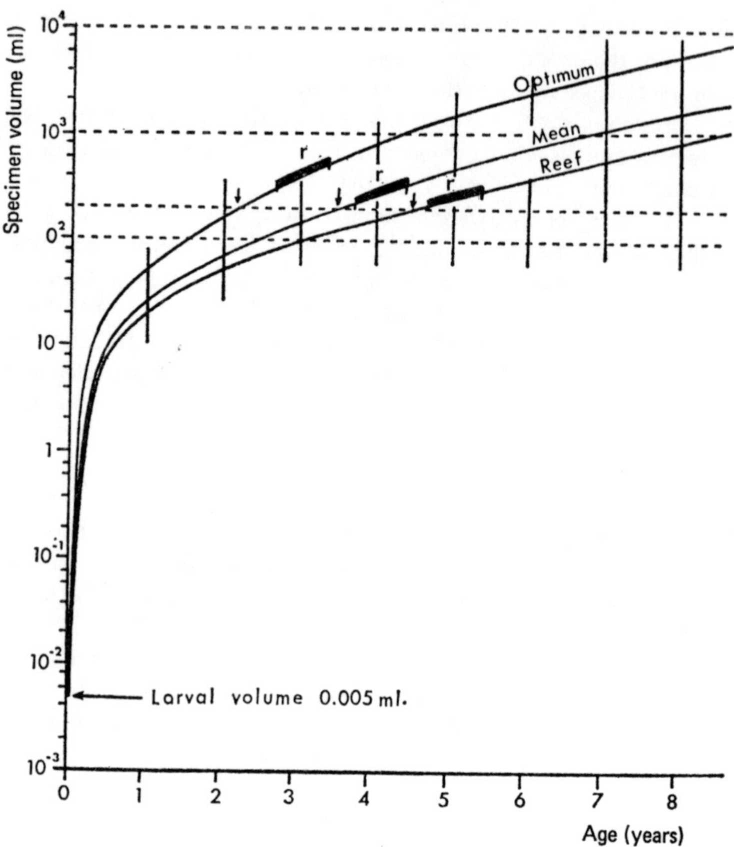

Fig. 6.2 Growth curves for *Mycale*.
Specimen size is plotted against age after settlement from mean growth rates of the solid-reef population, sand and reef populations and optimum observed rates. First reproductive season after reaching 200 ml volume (arrow) is indicated by r. (After Reiswig 1973.)

followed from settlement of single larvae, to reproduction in which an individual produced thirty larvae, over a period of four months. This sponge is an annual on New Zealand shores (Bergquist and Sinclair 1973). Other sponges show little growth over extended periods and appear not to reproduce annually. Specimens of such sponges, for example *Cliona* and *Ancorina*, are known which are more than twenty years old.

Reiswig (1973) has monitored growth in a subtidal population of *Mycale* and has discerned a clear pattern. Small individuals are subject to severe mortality as a result of biotic competition during the first year after settlement. When the sponge reaches 50–100 ml body volume, in one to three years depending on precise habitat (Figs 6.1, 6.2), competitors have little further effect. By the time they reach reproductive maturity in the third to the fifth year the sponges have assumed an upright tubular shape and reached a body volume of around 200 ml. Subsequent growth slows to a 60 % increase in volume each year, probably as a result of channelling energy resources into gamete and larval production. Since in the Jamaican sand and reef habitat 95 % of the standing crop of the population consists of specimens over 200 ml net volume, the effective growth rate of the population can be estimated at 60 % per year. It should be noted that this particular species of *Mycale* suffers high percentage mortality, particularly in response to storm conditions, It is an erect, fragile sponge with a small attachment base and thus is easily displaced. The emphasis on fast growth, huge larval production and high metabolic rate (see Chapter 1) always finds *Mycale* able to colonize new surfaces as they are available. It is not a sponge which survives long in a near static growth condition, as many do in space-limited conditions.

Sponges are also known to be able to enter into complex epizoic relationships, growing over, upon, or even inside one another without detriment to the pumping and filtering activities on which they depend (Fig. 6.3). Where different species settle on or overgrow each other, the attachment of the uppermost sponge to the basal component, which provides firm attachment, is usually undulating and loose, organized in such a way that the inhalant and exhalant system of the buried sponge can still operate. It has been suggested that only those sponges which have a reinforced ectosomal region within which inhalant and exhalant cavities can remain expanded while overgrown can function as the basal components for epizoic complexes. Certainly members of the tough Dictyoceratida and the long-lived, heavily siliceous, Choristida are involved frequently as the basis of epizoic structures. However, this generalization needs further examination.

Another avenue open to sponges which offers an escape from the effects of direct spatial competition is fusion. We have considered this in relation to

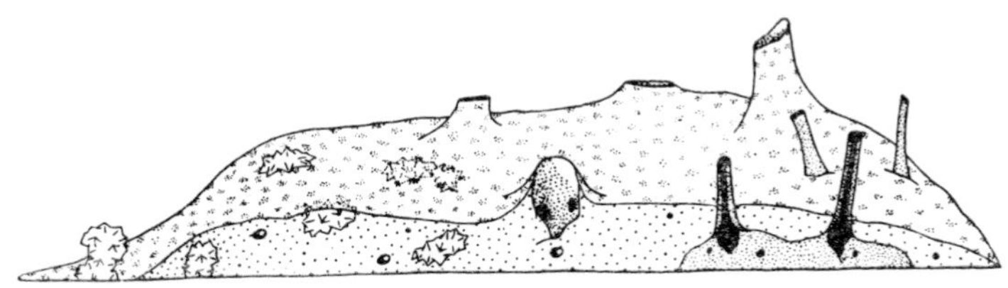

Fig. 6.3 A diagrammatic vertical section through a typical epizoic complex. *Gellius fibulatus* is supporting and overgrowing *Aplysilla sulphurea* (*left*) which also occurs completely enclosed. *Spongia virgultosa* (*right*) is overgrown by *Gellius* and only the oscular tubes show on the surface. (From Rützler 1970.)

Plate 9

(a) A typical sponge community of submerged rocky reefs in areas of strong water movement. In such areas the tough branching Axinellida and the large sperical Dictyoceratida are extremely common.

In this view the sponges are *Axinella* (branching) and *Fasciospongia* (globular with two oscular tubes). (Photo W. Doak.)

(b) *Polymastia* (Demospongiae: Hadromerida). A sponge with differentiated inhalant (short) and exhalant (long) surface papillae. (Photo W. Doak.)

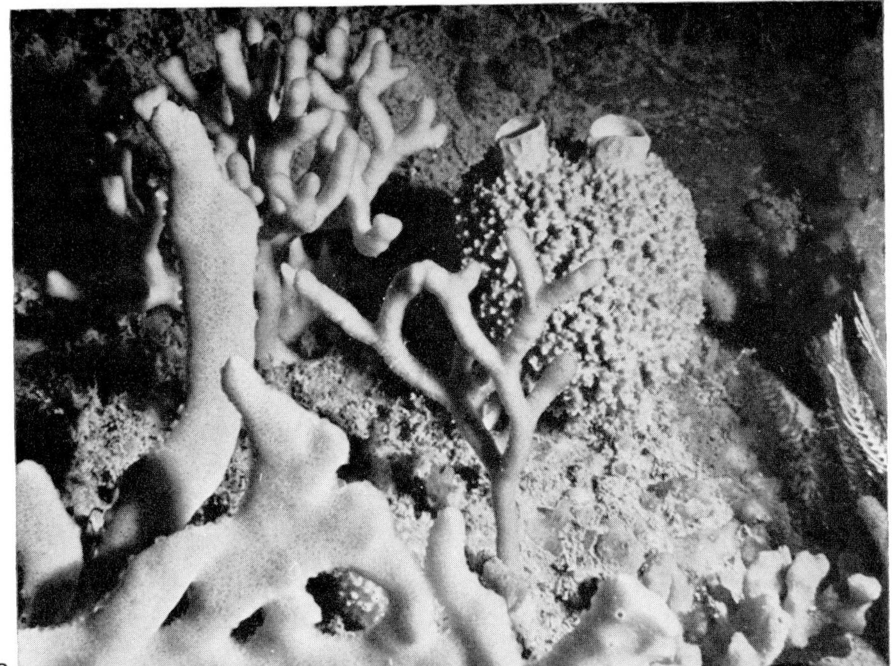
a

b

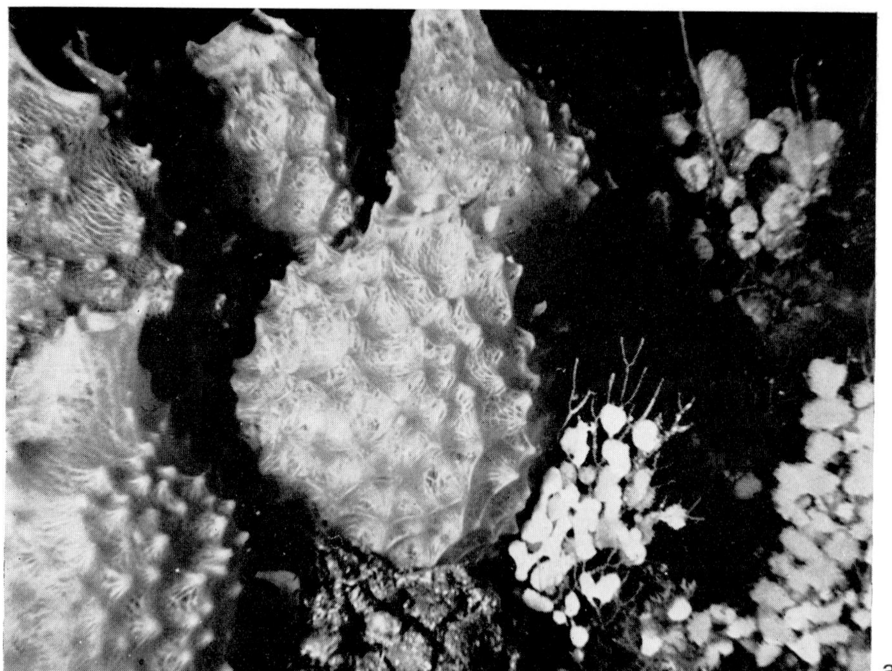

a

b

Plate 10

(a) *Tethya* (Demospongiae: Hadromerida). A group of individuals growing on shallow submerged rock. The elevation of the sponge surface into conules where the spicule tracts raise the fine surface membrane is well shown. (Photo R. Grace.)

(b) **Cliona celata** (Demospongiae: Hadromerida). A sponge which bores into calcareous material and then overgrows it to form a massive individual. The differentiated oscular (large) and ostial (small) apertures are obvious.

The branching sponge is *Trachycladus* (Demospongiae: Axinellida). (Photo W. Doak.)

larvae, but adult sponges of the same species, or sometimes the same physiological strain, coming into contact can also fuse. This cannot be construed simply as a mechanism which avoids competition, since it creates no space. It is advantageous to a less fit competitor which would otherwise be eliminated. It is likely to be most frequent between members of a clone. Implicit in this is the existence of finely attuned self-recognition mechanisms. These were possibly evolved because, in organisms such as sponges or colonial ascidians which have ill-defined individuality, members of one clone are interspersed with members of another and also with other species in space-limited habitats. It is very possible that organisms are competing for space with others which are effectively 'self'. Fusion under these circumstances is a way to greatly increase body size, hold the space, enhance reproductive potential, or reach reproductive mass more quickly.

The possibility that sponges of the same strain or species can recognize self and on this chemical basis can either fuse, establish static zones of non-coalescence, or overgrow a neighbour, brings us to the last significant point of sponge biology which is pertinent to their interactions with other organisms – their biochemistry.

Sponges are known to produce diffusible surface-active molecules which operate to promote or prevent cell-to-cell adhesion. Such factors, which are strain or species specific, were discussed in Chapter 2. A simple system of self-recognition, which leads to fusion or non-fusion, could be established as a result of the operation of factors of this type. Their existence has been well authenticated in a wide range of sponges. Such a system could be of some importance in sponge–sponge interactions in a space-limited community, but would be of no significance in interactions involving sponges from widely different systematic groups or involving sponges and other organisms.

However, in this broader sphere of spatial interactions, another biochemical attribute of sponges is certainly significant. That is the widespread production by sponges of compounds which at low concentration are toxic to other organisms. Toxicity is sometimes overt, as in the case of many species of *Tedania* and *Neofibularia* which produce an irritant mucus which can affect man; generally, however, the effect is not that spectacular. Clearly such chemical compounds afford some protection against overgrowth, either by killing the competitors, or by effectively rendering the substrate around the sponge or the sponge surface itself unsuitable for the settlement of larvae of other organisms. The widespread occurrence of antibacterial agents (Chapter 7) in sponges is likely to be more significant in the context of sponge feeding than in any competitive interaction with other organisms.

Jackson and Buss (1975) have recently published the first detailed ecological study that takes some of the above sponge attributes into account, and their results, based on observations on West Indian coral reefs, are worth considering in a little detail.

They observed that ectoproct colonies which were being overgrown by sponges frequently had a band of dead zooids around the sponge border. This suggested that allelochemical or more simply overt biochemical effects could be operating in what they termed 'interference competition'. To test for this they prepared aqueous homogenates of nine sponges and two colonial ascidians and assayed the effects these solutions had on the survival, or the activity, of three ectoprocts, two serpulids, a brachiopod and a bivalve. Extracts of five of the sponges and one of the ascidians exerted some allelochemical effects. In all of these cases, mortality, or cessation of movement or feeding, occurred for all replicates. The control animals suffered no mortality or abnormal behaviour. The allelochemical effects appeared to be quite specific among the ectoprocts, as no sponge or asicidian caused mortality or inactivation in all ectoproct species tested. Solitary animals were not affected at all.

While such experiments do not prove allelochemical effects *in vivo*, nor, if one assumes they are present, do they provide any indication of how the toxic mechanisms actually work, they certainly provide a strong suggestion that spatial competition can be influenced by biochemical mechanisms and that sponges are active organisms in this regard.

Ecological observations further support this contention. On the well-studied Jamaican fore-reef slope, assessment of sponge–coral interactions indicates that sponges overgrow corals much more frequently than the reverse, although when the reverse occurs, the sponge tissue shows no adverse effect. Bare zones, 1–3 mm wide, of dead coral skeletons adjacent to sponges are common. In this connection excessive mucus production by the sponge could serve to concentrate the allelochemicals at the sponge surface, although Bergquist and Bedford (in press) found no correlation between the incidence of antimicrobial activity and pronounced mucus production in a study of twenty-five species of temperate zone sponges.

How the possession of allelochemical agents might relate to spatial competition can be seen when attention is paid to the precise location and growth behaviour of particular sets of species. For example, two of the ectoproct species tested by Jackson and Buss (*Stylopoma spongites* and *Steganoporella magnilibris*) are the most successful ectoprocts under foliose corals. Both species will overgrow encrusting sponges. More frequently, however, they are overgrown by the sponges. Both ectoprocts have growth patterns that raise the growing edge or the feeding surface off the substrate. This habit should inhibit overgrowth and permit the ectoproct to overshadow prostrate organisms. The use of allelochemicals against such potentially important space competitors could be a major mechanism in preventing their dominance of the cryptic reef habitat. Sponges remain dominant in this habitat. They cannot physically overgrow such competitors, but they can, and do, inhibit them biochemically.

It is not surprising that there was no toxic effect of sponge homogenates against solitary animals. In the cryptic habitat these organisms are usually overgrown, eventually completely, but they occupy little space. The area occupied by solitary animal feeding apertures was less than 0.1 % of the total space available in the reef habitat studied by Jackson and Buss. There would be little selective advantage to the evolution of specific allelochemicals for eliminating such minor irritants.

This analysis of interactions involving growth form, rate, and allelochemical devices allowed its authors to postulate that complex competitive networks exist between the more than 300 encrusting invertebrate species in the cryptic reef habitat. Recall that we are discussing a high diversity, space-limited system in which physical disturbance and predation are minimal or absent. Instead of the overall competitive ability of space-occupying organisms following a linear hierarchy, where Species A > Species B > Species C > Species D, the system may be structured as a network, Species A > Species B > Species C > Species D, but Species D may win over Species A or B (Fig. 6.4). If Sponge 1 could overgrow Sponge 2 in all interactions and Sponge 2 was always toxic to ectoproct 1, while ectoproct 1 could always overgrow Sponge 1, then no clear dominant could emerge in this three-species system.

The more species involved and the more numerous and complex the networks, the slower will the space be occupied by a single competitive dominant, and less external disturbances will be required to maintain a given level of diversity in the system.

Clearly sponges possess allelochemical mechanisms which make them vigorous competitors in such habitats. The many reproductive devices they exhibit will contribute to the same vigour. To take just one example; some sponges instead of releasing larvae, either as a result of incubation or external development, incubate complete young sponges which then simply slide down the parent body and move onto the rooting stolons in the case of *Tethya*, or any adjacent vacant surface in the case of *Tetilla*. These have a very effective mechanism for ensuring

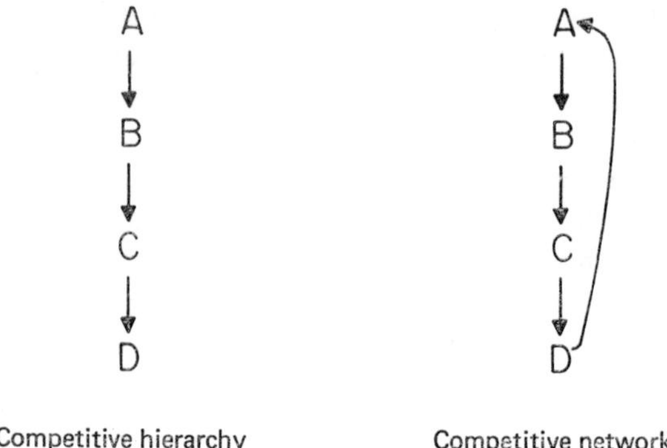

Fig. 6.4 A comparison of competitive networks and competitive hierarchies. Any species presented below another is the loser in competition with the species above. Arrows point from the dominant of a two-species interaction to the inferior competitor. (After Buss 1976.)

that new individuals of their own species obtain suitable living space. This mechanism is common in deep-water sponges, but is being observed with greater frequency in many habitats. Larvae which creep rather than swim could have the same effect.

With certainty, there is a fruitful area for future research in interpreting the various reproductive, physiological and biochemical strategies that have made sponges such successful organisms in colonizing submerged hard surfaces. Sound systematics, it hardly needs to be said, is the basis for any such study.

6.3.3 Sponges in biological associations with other organisms

(a) Higher invertebrates

Reference has been made earlier to the habitual association between certain sponges and some gastropods and bivalves. In these associations there is no suggestion that the mollusc has any other than a passive role, providing a suitable settlement surface for the sponge and possibly gaining a measure of camouflage. However, there are many associations involving sponges in which crabs effectively collect sponges and plant them on their backs. The sponge in some cases then moulds to the contours of the crab and provides a complete dorsal and lateral cover. Members of the Suberitidae are commonly involved in associations with the dromid 'camouflage crabs'. Spider crabs have the same habits, but are more cosmopolitan in their choice of sponges and can be found with three or four species sprouting from their backs.

Other well-known associations involve large sponges, usually Hexactinellida, Dictyoceratida or Hadromerida and many species of shrimps or worms. In these instances, the existence of a water current passing through the sponge body satisfies the nutritional and respiratory needs of the commensal and also conveys chemical or hydrodynamic information which guides the larval stages to the host. A single specimen of *Spheciospongia vesparia* from Florida was found to contain over 16000 shrimps belonging to the genus *Synalpheus*.

Another common association which obviously can be highly specific, but on which little study has been done, is that between species of Zoanthidea and sponges of many groups. One species of *Epizoanthus* colonizes the huge siliceous spicules making up the rooting tuft of *Hyalonema*; other genera are common as either surface colonizers or seated into inhalant apertures in sponges such as *Spirastrella, Verongia* and *Agelas*.

(b) Unicellular organisms

A most spectacular association is that seen between sponges and bacteria. In members of the Verongida the population of bacteria within the sponge mesohyl accounts for as much as 38 % of the tissue volume (Bertrand and Vacelet 1971), almost equal to the mesohyl volume (41 %) and almost twice as much as the actual cellular volume (21 %). Predominantly these bacteria belong to the genera *Pseudomonas* and *Aeromonas*. It is quite striking to note that, while the

Verongida in particular, and to a lesser extent the Dictyoceratida, support a high matrix bacterial population, many sponges have almost no matrix bacteria, for example, members of the Halichondrida and Haplosclerida. The reasons for this difference are not known, nor is it well known what exchanges take place between the bacteria and their sponge hosts. Certainly the sponge phagocytoses the bacteria, otherwise there would be no control over this population in which many dividing bacteria can be seen (Pl. 4*a*). In a study concerned with screening for antibacterial activities in marine sponges (Bergquist and Bedford, in press), we have noted a frequent growth enhancement in our test bacteria when provided with homogenates of the sponge matrix. This effect signifies that the sponge matrix affords a rich medium for bacterial growth, certainly enhancing growth over normal sea-water conditions. Vacelet (1975) has recorded the transfer of matrix bacteria in the larvae of Dictyoceratida. Thus the young sponge begins life with its bacterial flora established. This association appears to be a mutualistic symbiosis, where the sponge can phagocytose the bacteria and the bacteria are provided with a medium which enhances their growth.

Another remarkable series of associations is developed between sponges and various Cyanophyceae, the blue-green algae. Sponges are in fact the only metazoans which live in symbiotic associations with these organisms. The algae, like the bacteria, are mainly extracellular and often appear to compete with bacteria for matrix space. In *Verongia aerophoba* the superficial regions of the sponge contain many blue-green algae and the bacteria are excluded to deeper regions. This is not, as has been suggested, a response to production of antibacterial agents by the algae, since in the closely allied species *Verongia cavernicola* the same algae and bacteria co-habit and the level and chemical nature of antibiotic production by the two sponges is identical.

There are many observations indicating that the Cyanophyceae are phagocytosed by sponges. Their glycogen reserves have been identified in sponge cells (Sara 1971), and could of course be utilized by the sponge.

Associations between sponges and zoochlorellae are common in the freshwater spongillids and have been well studied. In marine sponges the association with zooxanthellae appears less frequent; they occur notably in species of the genus *Cliona*. The nature of the association probably parallels that existing between zooxanthellae and corals. The algae are not phagocytosed, but, in response to signals from the host, release photosynthetic products and other metabolites (Sara and Vacelet 1973).

6.4 Boring sponges: their role in bioerosion

Sponges which are able to excavate complex galleries in calcareous material belong predominantly to one family of the Demospongiae, the Clionidae, although some members of both the Spirastrellidae and the Adociidae can also bore. Many marine organisms are able to erode limestones and among these the sponges are of special interest because of the efficient means of substratum penetration by cellular etching that they have developed. The mechanism by which the sponge bores into the substrate has been much studied, not only because of its intrinsic interest, but because the activities of boring sponges inflict severe losses on commercial oyster fisheries, as they make the shells fragile and likely to shatter on opening.

During penetration by the sponge, the calcareous material is gradually destroyed as the sponge hollows out an extensive system of cavities and tunnels. These excavations are extended as small fragments of calcareous material are removed by the activity of specialized archaeocytes termed 'etching cells', and then expelled in the exhalant stream. Penetration of the substrate occurs along the interface where these cells contact the surface and the pattern of etching is thus quite complex. Each active cell releases a substance, possibly carbonic anhydrase, which dissolves the calcareous material around its edge to form a linear etching which corresponds to the contours of the cell. Deeper etching

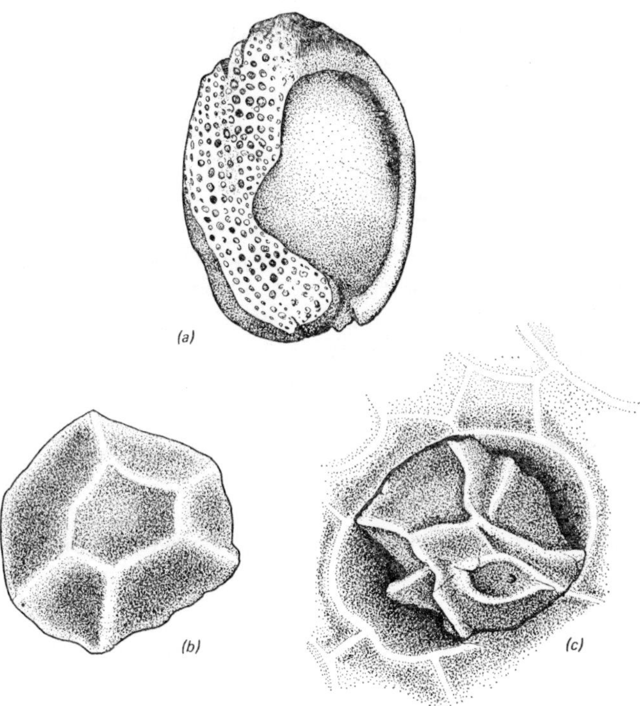

Fig. 6.5 The activity of boring sponges.
(*a*) A section of lamellibranch shell cut to show the galleries excavated by clionids; (*b*) a single carbonate chip which has been excavated by a sponge; (*c*) a similar chip still *in situ* to show how the cellular processes must surround the chip before removal. (Redrawn from Rützler 1975.)

occurs as the cell margins move downward through the initial groove and sink into the substrate in a noose-like fashion. During the sinking, the cell border is drawn down through the crevice cut by the cell edge, while the nucleus remains in position on the surface of the substrate within the original etched outline. Finally, when undercutting is completed, the small multifaceted chip is detached and expelled. Thus penetration is achieved by the precise cellular release of a chemical agent which dissolves the calcareous substrate along restricted zones of contact with specialized cells (Fig. 6.5).

The production of discrete calcium carbonate chips by boring sponges allows their detection in sediments and thus permits estimates of the activities of sponges in the erosion of coral reefs not only at present, but also in a palaeoecological context.

Following laboratory experiments, Neumann (1966) estimated that *Cliona lampa*, a common boring sponge on Bermuda reefs, was capable of removing as much as 6–7 kg of material from 1 m^2 of calcareous substrate in 100 days. Very little of the removal of substrate can be attributed to chemical dissolution, the most recent estimate being around 2–3 % (Rützler and Rieger 1973). Therefore, this means that between 5 and 6 kg of fine carbonate detritus can be generated from 1 m^2 of sponge-infected substrate in 100 days.

Neumann monitored activity only over 100 days. In order to assess in geological terms the importance of boring sponges in attrition of coral substrates, it is necessary to determine that this level of activity persists throughout the life of the sponge. Rützler (1975) has demonstrated that, indeed, initial erosive activity is significantly higher than long-term activity in *C. lampa*. His conclusion is, that after an initial adaptation period of one month, the rate curve of erosion in relation to actual surface covered by the sponge is steep, but flattens after six months (Fig. 6.6). After this time, the stimulation of the new substrate has been surpassed by competition for space and food which may retard boring activity. Energy will also need to be diverted from burrowing to nutrient storage and reproduction. It is interesting to note that in Rützler's experiments, when about 50 % of the available substrate was removed and the experimental blocks almost overgrown, the burrowing activity slowed almost to a stop (refer to the low mean value at seven and ten months in Fig. 6.6). This appears to hold for the behaviour of *C. lampa* both in the field and in the laboratory, while other species such as *C. celata* and *C. viridis*, destroy their substrate absolutely, and grow to become massive sponges.

The growth stages of clionids are referred to as alpha stages when the inhalant and exhalant performations extending from the eroded substrate are all that is visible of the sponge. Beta stages are seen in sponges where sponge tissue covers the surface of the substrate and links up the pores and oscules. The massive gamma stages are attained by only a few clionid species.

On the basis of Rützler's analysis of long-term boring activity, a mean rate of 700 mg cm^{-2} year^{-1} of substrate removal was estimated for *C. lampa*. If these calculations are extended they convert to 16 mg $CaCO_3$ mg^{-1} sponge dry

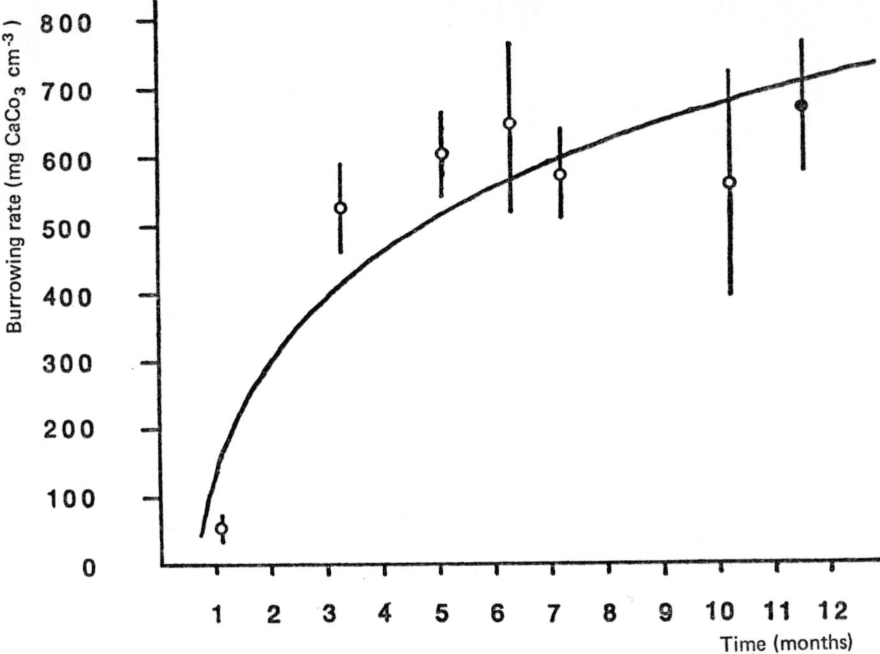

Fig. 6.6 The burrowing rates of *Cliona lampa* (○) and *C. aprica* (●) expressed as mg $CaCO_3$ cm^{-2} sponge in relation to final sponge surface area. Data points represent means plus or minus the standard error. (After Rützler 1975.)

weight $year^{-1}$. Thus sponges could be responsible for the erosion of up to 256 g m^{-2} $year^{-1}$ on the average in Bermuda. This would remove a rock layer 0.1 mm thick. In areas of dense sponge infestation, this figure can rise to 3 kg m^{-2} $year^{-1}$ equal to 1 mm depth of reef per year. This activity, Neumann suggests, has created a marked subtidal notch on Bermuda reefs.

Quite apart from changing the topography of reefs, clionids have a marked effect on the biotic interactions on submerged reef surfaces. Goreau and Hartman (1963) drew attention to the fact that clionid action in weakening the attachment of the huge foliose corals of the Jamaican fore-reef slope is directly responsible for the avalanches which ensue when these are torn off by storms, or dislodged by their own weight. The loss of these longer lived potential frame builders of the reef means that below 30 m depth, where clionid density is high, no reef framework is developed.

7 Sponge biochemistry

7.1 Introduction

There is great interest at present in the biochemistry of marine invertebrate organisms. This has been stimulated mainly by the recognition that the search for natural products with new molecular structures is likely to prove rewarding in this little-researched area. There is now a considerable literature indicating that not only are marine invertebrates a rich source of novel chemical structures (Baker and Murphy-Steinmann 1976), but that many of the naturally occurring compounds exhibit what is termed biological activity. Biological activity covers a whole range of effects such as the inhibition of the growth of micro-organisms, toxic effects against other marine organisms, and all pharmacological effects. This makes marine organisms attractive source material for the pharmaceutical industry, which must continually search for new chemical structures which are active in a multitude of ways. Another significant aim in the search for novel compounds has been to find molecules which can be modified chemically to produce analogues of known structures. Effectively nature does the difficult part of the synthesis, and the chemist does the rest. This area has proved very rewarding in several instances, notably in the discovery of prostaglandin precursors in gorgonians (Weinheimer and Spraggins 1969) and of the arabinose nucleosides in sponges (Bergmann and Feeney 1950).

While the interest of the chemist is in novel structures and the interpretation of their biosynthesis, and that of the drug industry lies in finding, identifying and synthesizing compounds with useful activity, the biochemical diversity which continues to be revealed in some groups of marine invertebrates also holds many interesting and rewarding avenues of research for the biologist.

Historically, sponges were the first marine invertebrate group to be studied in any comprehensive way by those searching for new compounds, and up to the present, they have yielded a great number of novel compounds over a very diverse chemical spectrum. What can the sponge biologist take from this growing body of data? First, consider possible reasons for the demonstrated chemical diversity. Sponges are an ancient group of invertebrates; the geological history of all classes stretches back at least to Cambrian time. They are also successful organisms in their many habitats. In the course of this long evolutionary history, it is certain that many metabolic, physiological and ecological alternatives have been developed within the group. The rich array of unusual chemical compounds which are often indicators of novel biosynthetic pathways, are the traces that remain of these diverse evolutionary experiments. Analysis of particular molecular species across a representative selection of sponges could possibly provide some comparative measure of evolutionary separation between taxa in terms of actual molecular differences. In this connection comparisons of DNA and some enzymes would be rewarding, since these could be compared with an existing body of information for other invertebrate groups and thus provide information on the degree to which sponges and other metazoans are related. Some analyses of this type are available, but as yet none that deal with an informative selection of sponges. This lies in the future.

Another avenue of research is to turn the biochemical diversity of sponges to advantage in systematics. The presence or absence, and biosynthetic relationships, of particular compounds can provide additional characteristics which are useful, both when attempting to evaluate existing classifications, and when seeking to establish new taxonomic arrangements. A biochemical approach to taxonomy is particularly pertinent when applied to sponges where the macroscopic features of the organism provide few clues to their real relationships. There is a huge potential body of information which has application to systematics in the comparative biochemistry of sponges. To date, however, there have been very few papers which can be considered to be contributions to the biochemical taxonomy of the Porifera. It is essential in any such study that a sufficient number of sponges are included to enable group characteristics to be recognized and to establish the levels of in-group variability with respect to the compounds being compared. Further, it is essential that the identification of the sponges be accurate; poor identification renders the work useless, and, because the errors are frequently perpetuated in review literature, it can be actively misleading. For example, a major review of sponge chemistry and intermediary metabolism (Hammen and Florkin 1968) records two species of the Demospongiae as belonging to the Calcarea; Jacobson and Smith (1968) in reviewing the biochemistry of taurine and its derivatives follow Kittredge, Simonsen, Roberts and Jelinek (1962) in referring *Xestospongia* (Demospongiae) to the Calcarea. Biochemical taxonomy cannot rest on such a basis. The solution to the problem lies in close co-operation between biochemists and systematists and in the active involvement of the latter in framing the studies, asking the questions and gathering the information. Papers which satisfy the above criteria are very few at present, but Bergmann's pioneering contributions dealing with sterols (Bergmann 1949, 1962) should be mentioned and more recently those of Bergquist and Hogg (1969) and Bergquist and Hartman (1969) dealing with free amino acids. Cimino, de Stefano, Minale and Sodano (1975) have been innovative in considering several types of compounds and relating the overall pattern to sponge taxonomy.

Another major area of general biological interest concerns the high incidence in sponges of compounds with biological activity as evidenced either by toxic or antimicrobial effects. It is obvious that the rationale for the occurrence of such compounds lies in the biology and ecology of the organisms. The expression of the activity in any pharmaceutical test situation is quite secondary. Recently the concept of allelochemical interactions discussed in the previous chapter has been framed (Whittaker and Feeney 1971). The realization that such biochemical defence mechanisms are subtle and widespread in the plant and animal kingdom has led logically to the idea that defence from predation, or effectiveness in competition, might involve the toxic compounds found in such profusion in marine invertebrates. In the case of antibacterial compounds it is most likely that they have a primary physiological role and only a secondary offensive or defensive one. Sponges retain bacteria in the water current with high efficiency (see Chapter 1). The presence of diffusible anti-bacterial agents in the matrix could contribute to this retention, causing clumping of the bacteria by the time they reach the choanocyte collar mucus filter, and thus enhancing the retention efficiency as a result of increasing the particle size.

It is beyond our scope to detail all the novel features of sponge biochemistry, but we can consider some groups of compounds and thus illustrate the types of research available to the biologist who may wish to exploit the biochemical diversity of the Porifera.

7.2 Some examples of the biochemical diversity of sponges

7.2.1 Sponge nucleosides and nucleic acids

The isolation of two unusual nucleosides from the Jamaican sponge *Tethya crypta* (Bergmann and Feeney 1950) was, perhaps more than any other single discovery, responsible for stimulating wide interest in sponges as a potential source of new chemical compounds. These nucleosides, which occurred free in the sponge rather than as components of the nucleic acids, were given the trivial names, spongouridine and spongothymidine. They are the 1-β-D-arabinosyl derivatives of the pyrimidines uracil and thymine respectively. The period following Bergmann's discovery saw a great growth of interest in nucleic acids

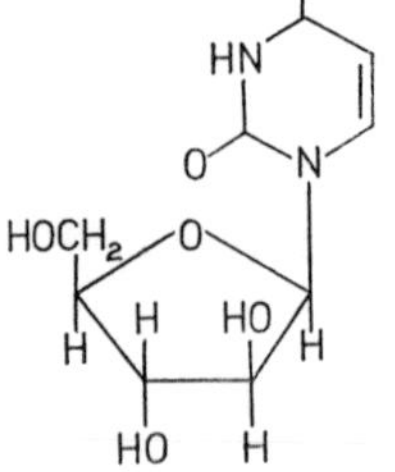

Spongouridine

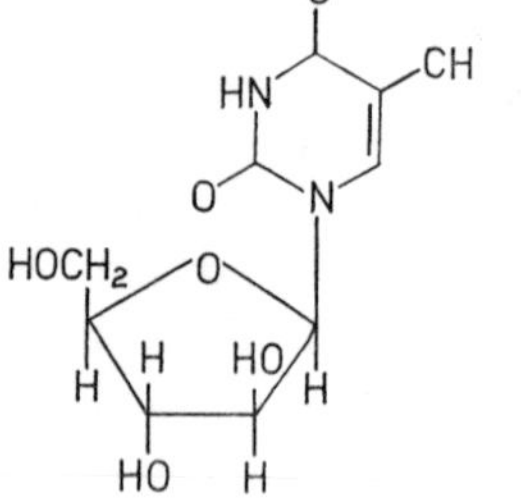

Spongothymidine

and their component molecules from many points of view, notable among which was the possibility of synthesizing chemical analogues which might possess anti-tumour activity. Spongouridine and spongothymidine did indeed serve as models for the synthesis of D-arabinosylcytosine, a nucleoside analogue which has antiviral and other biomedical properties which are attributable to its ability to inhibit key pathways in nucleic acid biosynthesis (Cohen 1963).

The fact that *T. crypta* possessed free arabinose nucleosides led biochemists to speculate as to the metabolic conditions under which the sponge tissue was operating. *Tethya crypta* is a species which lives in shallow water almost buried in fine sand with only the oscular complex elevated. Cohen (1963) postulated that *T. crypta* operated to some degree anaerobically since, compared to model bacterial systems, it was clear that the sponge had major difficulty in synthesizing long RNA chains and completing them with adenylic acid. In *Escherichia coli*, adenosine represents 80–90 % of the terminal nucleoside in RNA, while in *T. crypta* it represents only 30 %. This, coupled with the high concentration of free arabinosyl pyrimidine, is compatible with partial anaerobiosis. In this connection, it is interesting to note that Bergmann, Watkins and Stempien (1957) recorded an exceptionally low total RNA content in *T. crypta* in comparison to that of fifteen other sponges. These authors suggested the existence in this sponge of an unusual metabolic pathway which had the effect of diverting most of the normal intermediates in RNA synthesis to the formation of free spongouridine and spongothymidine. If this speculation linking the occurrence of free arabinose nucleosides with sponge tissue operating under conditions of partial anaerobiosis is valid then these compounds are likely to occur in a variety of sponges: for example *Coelocarteria singaporense, Ciocalypta penicillus, Spirastrella vagabunda, Cinachyra australiensis*, to name only a few.

So far these compounds are known only from *Tethya crypta*, but the species suggested above which have a similar habitat have never been investigated. It should be noted that there is no proof that *T. crypta*, although partially buried, does metabolize anaerobically. The diversity and quantity of sterols in *T. crypta* and in all of the above sponges suggest the reverse, since sterol biosynthesis is absolutely oxygen dependent.

7.2.2 Free amino acids

There are two quite different approaches which utilize comparative biochemistry to resolve taxonomic problems. It is possible to look for unique molecules in different taxa and then, if possible, to interpret the biosynthetic relationships existing between the compounds. Another approach is to examine the relative quantitative distribution of common intermediary metabolites. The studies carried out by Bergquist and Hogg (1969) and Bergquist and Hartman (1969) on free amino acid patterns in the Demospongiae used the latter approach. These studies went some way towards establishing amino acid 'fingerprints', obtained using two-dimensional thin-layer electrophoresis and chromatography, which characterized the orders and some families of the Demospongiae.

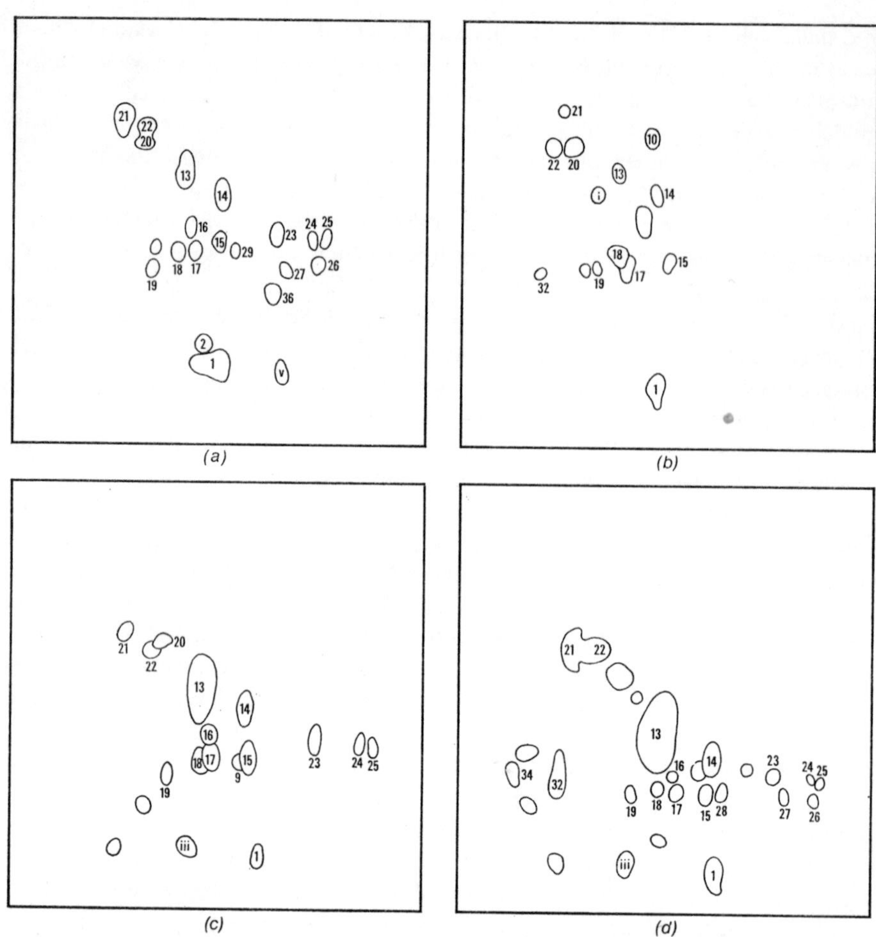

Fig. 7.1 Diagram of the free amino acid fingerprints of four species of Ceractinomorpha.
(*a*) *Gelliodes ramosa* (Haplosclerida); (*b*) *Verongia fistularis* (Verongida); (*c*) *Polyfibrospongia echina* (Dictyoceratida); (*d*) *Microciona prolifera* (Poecilosclerida). Numerals refer to the same ninhydrin-positive compounds throughout. Small roman numerals denote unidentified compounds. Unnumbered spots were not utilized for comparative purposes, and largely represent unidentified compounds. (1) Taurine. (2) Hypotaurine. (3) Taurocyamine. (4–7) (i–iv) unidentified compounds. (8) Sarcosine. (9) Pipecolic acid. (10) β-Alanine. (11) β-aminoisobutyric acid. (12) γ-Aminoisobutyric acid. (13) Glycine. (14) Alanine. (15) Threonine. (16) Serine. (17) Glutamine. (18) Glutamic acid. (19) Aspartic acid. (20) Arginine. (21) Lysine. (22) Histidine. (23) Valine. (24) Isoleucine. (25) Leucine. (26) Phenylalanine. (27) Methionine. (28) Methionine sulphoxide. (29) Methionine sulphone. (30) Proline. (31) Hydroxyproline. (32) Cysteine. (33) Tyrosine. (34) Asparagine. (35) Cysteic acid. (36) Tryptophan. (37) Ethanolamine. (38–44) (v–xi) unknown compounds.

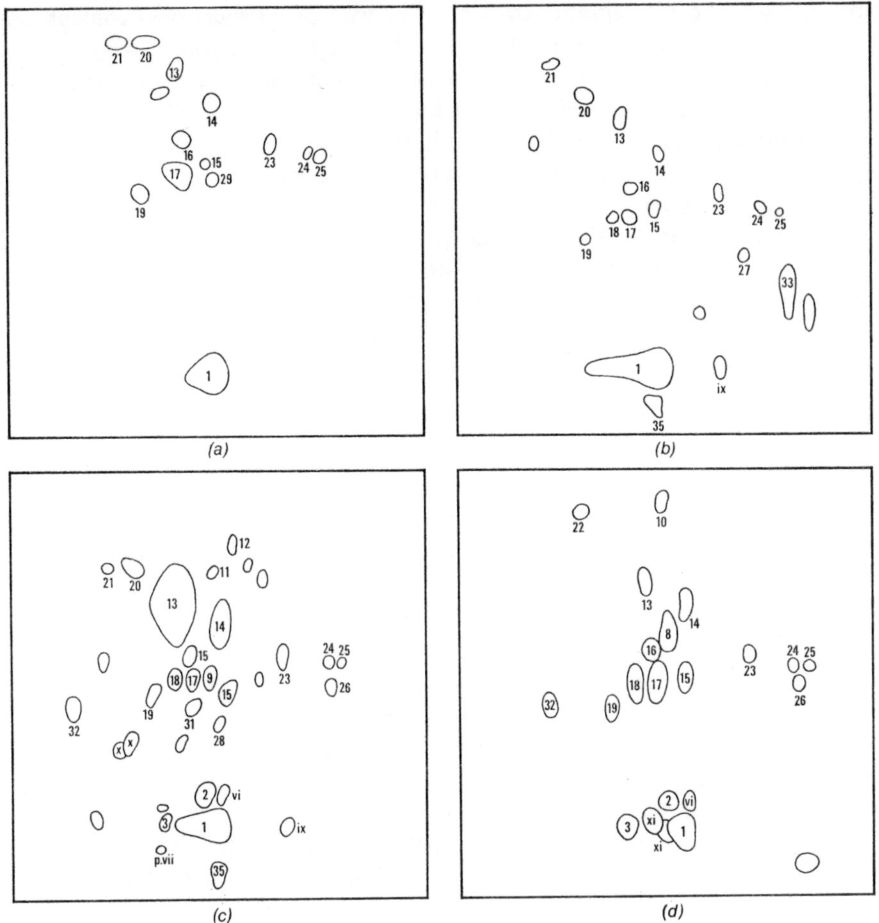

Fig. 7.2 Diagrams of the free amino acid fingerprints of four species of Tetractinomorpha.
(*a*) *Agelas dispar* (Axinellida); (*b*) *Homaxinella* sp. (Axinellida); (*c*) *Tethya* sp. (Hadromerida); (*d*) *Cinachyra* sp. (Spirophorida). (Notation as for the caption to Fig 7.1). (From Bergquist and Hartman 1969.)

A major goal of these studies was to evaluate the classification of the Demospongiae into two subclasses proposed by Levi (1956) mainly on the basis of reproductive biology. In the two publications the authors considered eighty-seven species of Demospongiae which included representatives of all orders. A major additional aim was to determine the level of classification at which variation, both qualitative and quantitative, in amino acid content could yield information of use to the taxonomist. Did patterns characterize species, genera, families, orders or subclasses? This is an important point, often lost on the biochemist. Not every unit character has equivalent informational content in a phylogenetic sense. All information must be weighed and considered.

The results of analysing free amino acid patterns were, very briefly, to support the homogeneity of the Ceractinomorpha with the reservation that the Verongiidae, at that time grouped within the family Spongiidae, were extremely divergent from other Ceractinomorpha (Fig. 7.1*b* as compared to Fig. 7.1*a*, *c*, *d*). Within the Ceractinomorpha, the Dictyoceratida and Poecilosclerida (Fig. 7.1*c*, *d*) were very similar, and the Haplosclerida (Fig. 7.1*a*) showed strong affinity to the family Halichondriidae within the Halichondrida. Within the Tetractinomorpha, every order displayed a characteristic pattern, underlining again the diversity of this assemblage when compared to the Ceractinomorpha (Fig. 7.2).

Two predictions which have now been vindicated by other biological information were made on the basis of this work. First, *Agelas*, long a 'problem' genus, showed most affinity with the Axinellida (Fig. 7.2*a*, *b*), a position now reinforced by reproductive information (Reiswig 1976); and, secondly, *Verongia* proved so different from other keratose or non-siliceous sponges, that it was separated at the family level. Other biochemical information, and the fact that the verongiids are oviparous, now supports this, and argues for even more distant separation.

It should be emphasized that in this study the authors had a comprehensive grasp of sponge systematics and designed a species sample which had the potential to answer the necessary questions. They concluded that some groups, notably the Choristida, were very conservative with respect to amino acid content, while others, the Spirophorida and Axinellida, were quite diverse. Clearly this particular biochemical attribute is worth pursuing where diversity was shown, but other approaches utilizing different molecular species could prove more rewarding where conservatism in amino acid pattern is shown. There is no reason to assume that all parameters change in evolution at the same rate in diverse groups; quite the contrary.

Some quite rare naturally occurring amino acids were recorded by Bergquist and Hartman (1969) including β-amino isobutyric acid and pipecolic acid.

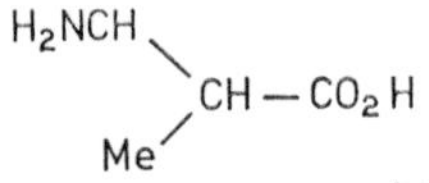

H_2NCH / $CH-CO_2H$ / Me

β-Amino isobutyric acid

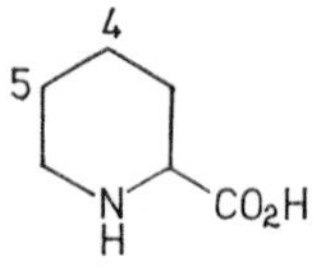

4, 5, N, H, CO_2H

Pipecolic acid

While Bergquist and Hartman were interested primarily in the ninhydrin-positive compounds, chiefly free amino acids, as components of a pattern which showed quantitative and qualitative variation from group to group, they also noted two points which could indicate the occurrence of differing metabolic pathways in particular groups.

Taurine is formed as the end-product of transulphuration from methionine, the compound which supplies organisms with most of their required sulphur. There are five demonstrated pathways by which the final steps of this transformation are effected (Jacobsen and Smith 1968); three involve hypotaurine as an intermediate and two do not. For example the most common pathway is methionine–cysteine–cystein–sulphuric acid–hypotaurine–taurine. The most likely alternative is methionine–cysteine–cystein sulphuric acid–cysteic acid–taurine. Most sponges contain hypotaurine (Figs. 7.1, 7.2), but it is lacking in four orders, the Poecilosclerida (Fig. 7.1*d*), Dictyoceratida (Fig. 7.1*c*), Verongida (Fig. 7.1*b*) and Dendroceratida and in several genera within the tetractinomorph assemblage. This could reflect the existence of different enzymatic pathways of taurine biosynthesis in different sponge groups. Another interesting consideration was that in sponges, as in some other invertebrates, taurine which often accumulates in extremely high quantities, is not necessarily a metabolic dead end. Ackermann and List (1959) have identified trimethyltaurine (taurobetaine) in *Geodia*, and Bergquist and Hartman (1969) and Roche and Robin (1954) reported the occurrence of taurocyamine (2-guanidoethane sulphinate) in many species of Hadromerida, Spirophorida and Choristida. Since taurocyamine functions as a phosphagen precursor, it would be interesting to know whether other more usual phosphagens also occur in those sponges which are capable of carrying out the transaminidation of taurine. Roche and Robin (1954) have reported the presence of phosphoarginine in *Hymeniacidon caruncula* and of phosphocreatine in *Tethya aurantium*. Further work on these compounds and their taurine-related precursors could help in characterizing high-level taxonomic categories within the Porifera.

7.2.3 Sterols

It has been mentioned earlier that the pioneer in the area of utilizing biochemical information in relation to sponge classification was Bergmann, who, in the 1940s and 1950s, with a sequence of collaborators investigated the fatty acids, sterols, nucleic acids and nucleosides of a number of sponges. Bergmann's major contribution, however, was his work on the structure and distribution of sponge sterols which is summarized in one publication (Bergmann 1949). His interest in sterol type in relation to sponge systematics was kindled by the finding that each of the first four sponges he studied yielded a novel and distinct sterol. Then, within what at that time was considered to be one family, he identified four distinct sterol compositions. This prompted him to comment: 'These astounding results of my preliminary survey encouraged further study; it was hoped we could provide data which could be incorporated into the taxonomy of sponges which is so bewildering to the uninitiated and understood by so few' (Bergmann 1962). Later, in the same review publication, he was to write 'after fifty sponges had been sampled the diversity was not as spectacular as had been first indicated'.

Since the 1950s the resolving power of analytical methods which can be applied to the separation of complex sterol mixtures has increased enormously. In fact many entities classed by Bergmann as single sterols have proved subsequently to be mixtures of two or three distinct compounds. These factors have contributed to a revival of interest in the sterols of sponges. De Rosa, Minale and Sodano (1973) have commented on the distribution of sterols in twenty-five sponges and a comprehensive biochemical taxonomic survey involving fifty-one species of Demospongiae representing thirty-six genera, nineteen families and ten of the eleven orders has recently been completed (Bergquist and Hofheinz, in press). This study has employed isolation by thin-layer chromatography followed by analysis using a gas chromatograph, mass spectrometer combination. Some analyses were supplemented by infra-red and nuclear magnetic resonance spectra. Unfortunately, many of Bergmann's data were reported only in outline, therefore it has not proved possible for later authors to use his work except in the most general terms to extend the scope of their own comparative studies.

The features of sterol biochemistry used by Bergquist and Hofheinz to examine systematic relationships were, first, the occurrence of any unsual sterols and then the number of different sterols present in a sponge. Further attention was paid to the size of the sterol molecule, in terms of the carbon number e.g. C_{26}, C_{27}, C_{28}, C_{29}, C_{30}, C_{31}, the type of tetracyclic ring skeleton, the location of any additional carbon, whether in the nucleus or on the side chain, the

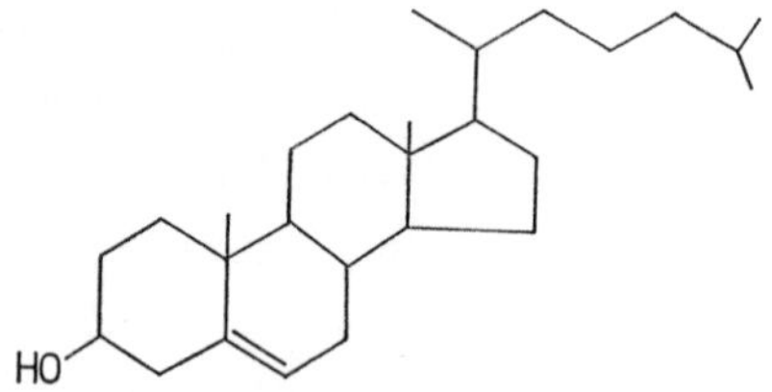

HO

Cholesterol (C_{27}, $\triangle 5$)

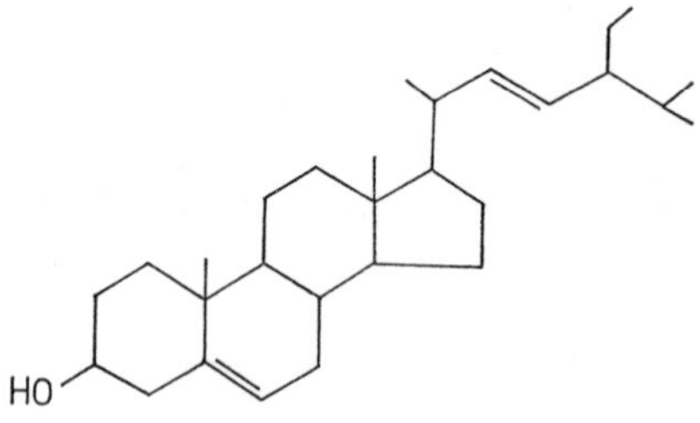

HO

Poriferasterol (C_{28}, $\triangle 5$)

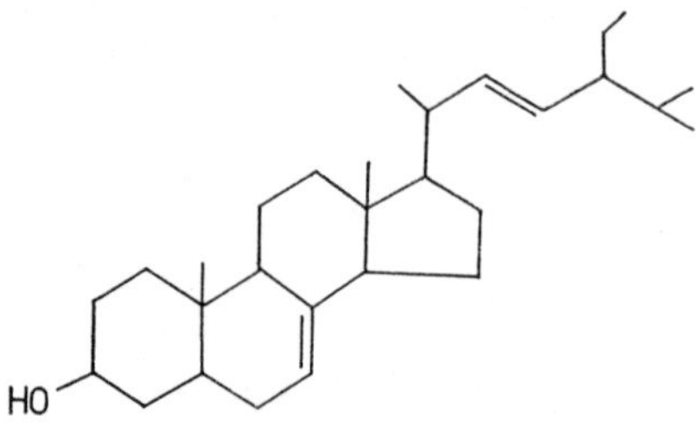

HO

Chondrillasterol (C_{28}, $\triangle 7$)

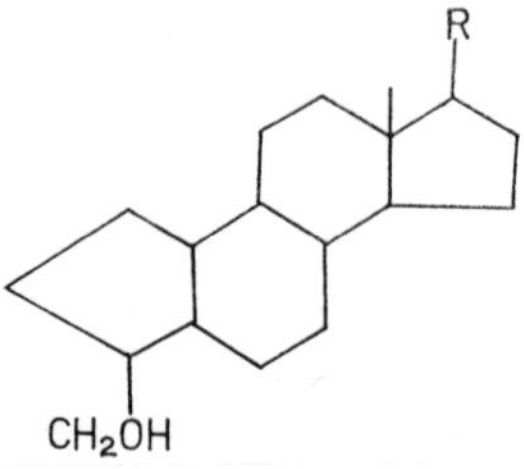

R

CH_2OH

A new type of norstanol from an axinellid sponge. It occurs in C_{27}, C_{28} and C_{29} forms.

location of double bonds, where present and in some cases the stereochemistry of the ring–side chain junction. This, plus some additional information from infra-red and nuclear magnetic resonance spectroscopy, gave a number of parameters by which the sterol mixtures could be compared. Gas chromatography without mass spectra provided percentage data for the components of all mixtures, and allowed comparison of overall sterol composition on the basis of the percentage of saturated to unsaturated compounds.

The results of this study have vindicated Bergmann's original supposition that amazing structural diversity was to be found in the sterols of the Demospongiae. Fifty-two distinct sterols were found, ranging in carbon content from C_{26} to C_{31}, the latter being the largest naturally occurring sterol recorded to date. The tetracyclic ring structure ranges from normal stanols (cholestanol) and sterols (cholesterol), to hydroxymethyl norstanols and Δ 5, 7 ketosterols, and double-bond configurations have been found in Δ 5, 7, 22, 24 (28) positions. The number of novel sterols, of which only a few were completely characterized, exceeded the number of previously recorded structures.

Within the Demospongiae there proved to be great variability in sterol content; some sponges possessed only two sterols (some Axinellida and Halichondria), others as many as sixteen (Haplosclerida), but most commonly there were between seven and ten distinct sterols present. In some species one sterol is present in high amounts and others only as traces. For example, in *Phyllospongia foliascens* cholesterol accounts for 82 % of the mixture, in other species, for example most Hadromerida, several sterols occur at about 15–20 % levels as determined by gas chromatography.

Sterols were found to occur at very variable concentrations, from less than 0.01–1.5 % of the sponge dry weight, between 0.5 and 50 % of the total lipids. Extremely low sterol content characterizes the Dictyoceratida and emphasizes once again the differences between these sponges and the Verongida which quantitatively have a normal sterol content. In the Dictyoceratida, relatively high amounts of terpenoid compounds are produced and may, in this order, represent a biosynthetic alternative to sterols. These terpenes usually show biological activity and will be considered later.

Bergquist and Hofheinz were able to affirm that sterol pattern is species specific and does not fluctuate markedly either in specimens of a single species collected from widely different localities or in a single species collected at different times in the same area.

Closely related species displayed similar sterol patterns; for example all the Verongiidae were very similar, likewise the four species of *Tethya* were very similar, but different lines within the genus, coincident with morphological distinctions, were indicated by different percentages of C_{27} and C_{28} sterols.

Striking within group similarity was found in the Dictyoceratida, Verongida Callyspongiidae and Hadromerida. Equally marked variation in sterol pattern was found within the Poecilosclerida, Halichondrida and Axinellida. In the case of the Poecilosclerida the species sample was too small to be representative of that huge order, but the diversity within the two latter groups is a real reflection of biochemical divergence, once again along lines similar to those suggested by some biological evidence. When the species sample worked on by Bergquist and Hofheinz is considered in conjunction with that used by De Rosa *et al.* (1973), it is possible to make some quite broadly based statements on the sterols of two groups in particular, the Axinellida and the Verongida.

Most members of the Verongida have as their major sterol components novel compounds with an aplystane skeleton (De Rosa *et al.* 1973). These compounds are very rare in other sponges and serve further to indicate the divergence of this group from the Dictyoceratida.

The Axinellida show extremely diverse patterns, and major differences in sterol structure, including the occurrence of new types of norstanols, characterize some species. It is obvious that the differences in sterol composition within and between families of the Axinellida are very great by comparison with those encountered elsewhere in the Demospongiae. This serves to emphasize the distinctness of this group from the more homogeneous Hadromerida and the unsatisfactory nature of the present family classification, in which all forms with a simple skeleton and no microscleres are lumped into the Axinellidae. Both De Rosa and Bergquist and Hofheinz included *Agelas oroides* from the Mediterranean in their sample; the latter authors also extracted *Agelas mauritania* from the Great Barrier Reef. There is an excellent coincidence of results. The sterol pattern for *Agelas* is constant in its major features within the genus (Table 7.1) and has much in common with one of the several patterns found within the Axinellida, specifically that shown in *Axinella damicornis*.

Table 7.1 *Comparison of the sterol composition of two species of* Agelas.

	Percentage of C_{27}	C_{28}	C_{29}	Main features
Agelas oroides (De Rosa *et al.*)	38	15	47	Stanols and $\Delta 7$ sterols
Agelas oroides (Bergquist and Hofheinz)	31	15	54	Stanols and $\Delta 7$ sterols
Agelas mauritiana (Bergquist and Hofheinz)	38	15	42	Stanols and $\Delta 7$ sterols

Finally, the study of comparative sterol composition reinforces the view expressed earlier in relation to free amino acid patterns, that with the exception of the Verongida, the subclass Ceractinomorpha is homogeneous and the subclass Tetractinomorpha is very heterogeneous. The biological role of these diverse sterols which are often present in high yield is absolutely unknown.

7.2.4 Compounds with biological activity

Many structurally quite diverse compounds which express biological activity have been isolated from sponges, but four groups are of particular interest. The terpenoids, the heavily halogenated dibromotyrosine-derived compounds, the bromopyrrole derivatives and the prenylated benzoquinones.

(a) Terpenoids

These compounds are mainly linear furans and have in the main been extracted from sponges belonging to the orders Dictyoceratida and Dendroceratida where clearly they are biosynthetic alternatives to the sterols. Some can be quite complex, for example, the sesterterpenes which require the linear combination of five isoprene units, and the polycyclic terpenoids such as scalarin.

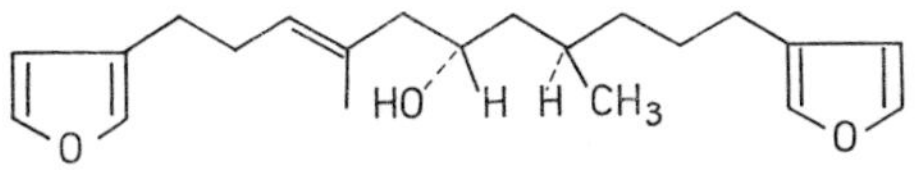

Furospongin–1, an antibiotic C-21 furantoterpene from *Spongia officinalis.*

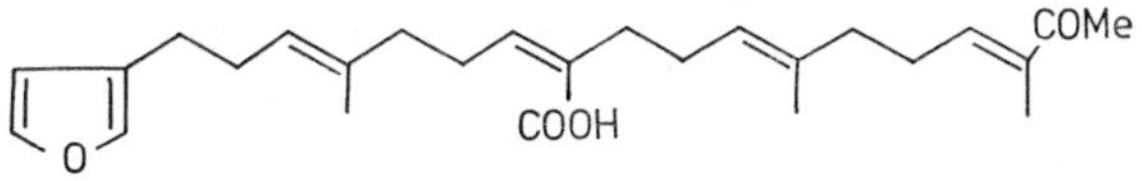

Furospongin–3, a sesterterpene from *Spongia officinalis.*

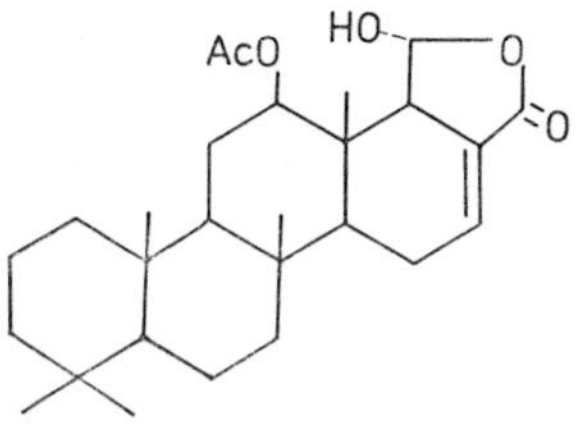

Scalarin, a cyclized sesterterpene from *Cacospongia scalaris.*

Most of the furanosesterterpenes, certainly all of those containing a tetronic acid unit, are strong antibiotic agents; for example ‘variabilin’ extracted from *Ircinia variabilis.*

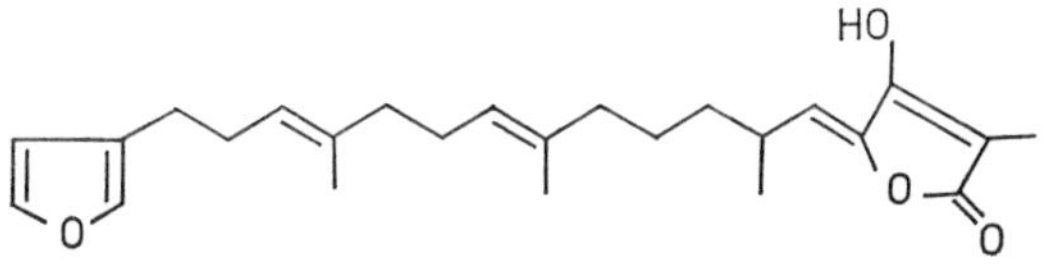

Variabilin.

Since these compounds are obviously related biosynthetically, and have now been found in many genera and species of the Dictyoceratida and Dendroceratida, there is reason to hope that they may in future provide material for further taxonomic manipulation. To date, however, the informational content of these compounds has not been evaluated except by chemists who find it difficult to interpret evolutionary taxonomy.

It is likely also, since terpenoid compounds are notably astringent to the taste, that their presence in high yield in sponge tissue causes the sponge to be unpalatable to selective predators. This could supplement the established antibiotic activity and be possibly more relevant than antibiotic activity in an allelochemical context. The biological properties of naturally occurring terpenoids deserve detailed investigation in relation to the ecology of the non-siliceous sponges.

(b) Bromopyrrole compounds

Bromopyrrole derivatives have been found in only five species (Cimino *et al.* 1975), two species of *Agelas*, two species of *Axinella* and *Phakellia flabellata*. The more complex members of this series of compounds include 'oroidin' derived from *Agelas oroides* from the Mediterranean and now known also to occur in *Axinella verrucosa* and *Axinella damicornis*. *Phakellia flabellata* from the Great Barrier Reef has yielded dibromophakellin, a polycyclic compound related

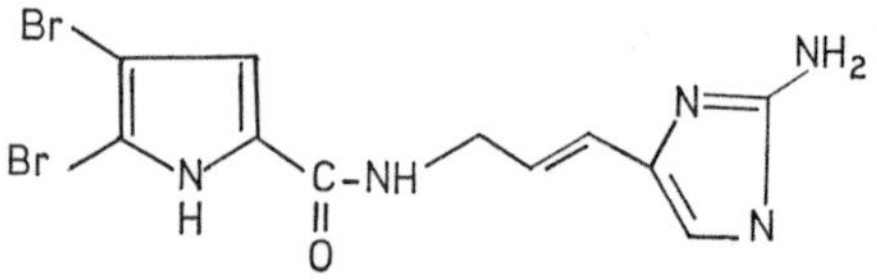

Oroidin, a bromo-pyrrole compound from *Agelas oroides*.

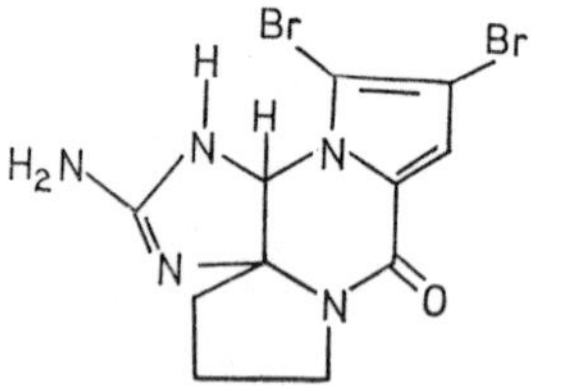

Dibromophakellin, a polycyclic bromo-pyrrole derivative from *Phakellia flabellata*.

biosynthetically to oroidin. The occurrence of these unusual, related metabolites in *Axinella*, *Phakellia* and *Agelas* provides one further indication that, as suggested initially by Bergquist and Hartman (1969), *Agelas* is related to the Axinellida rather than to the Poecilosclerida.

(c) Compounds derived from dibromotyrosine

These are a series of small molecules, all containing bromine, some of which are responsible for the marked antibiotic activity shown by all species of the Verongida thus far investigated. They appear to be restricted in occurrence to

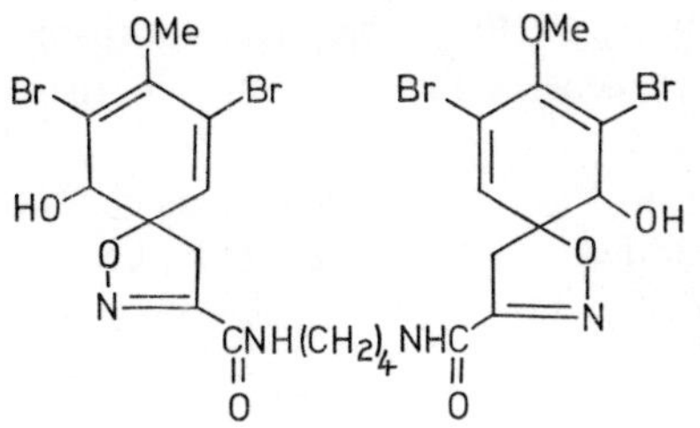

Aerothionin, a novel tetrabromo compound with antiobotic activity from *Verongia aerophoba.*

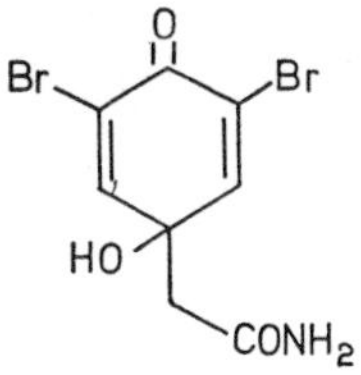

A related dibromo compound from *Verongia fistularis.*

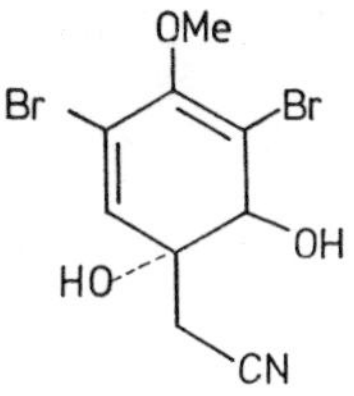

Aeroplysinin–1 from *Verongia aerophoba.*

the Verongida (Cimino *et al.* 1975), and interpretation of their biosynthesis may possibly shed light on the problem of generic definition within this order where present divisions rest mainly on the quantity and structure of the fibres, and the comparative histology of the secretory cells.

(d) Prenylated benzoquinones

Two species of *Ircinia* have yielded a quite novel group of prenylated benzoquinones and the corresponding quinols (Cimino, de Stefano and Minale 1972)

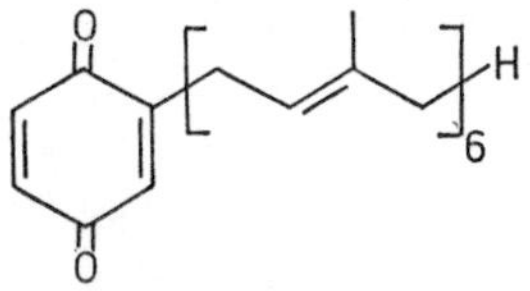

A prenylated benzoquinone from *Ircinia spinulosa.*

These compounds are related to chemicals which in other invertebrate groups function to confuse the olfactory sense of predators (Kittredge, Takahashi, Lindsey and Laska 1974). This function has not been demonstrated for the sponge quinones, but further research on this type of activity could prove rewarding.

7.2.5 Sponge pigments

Spectacular and diverse colouration is one of the most obvious features of sponges. Despite this, there has been little work directed specifically towards the chemistry of sponge pigments. The lack of chemical literature, coupled with the absence of simple routine but quantitative methods for extraction and characterization of pigments, means that in general the chemistry of these compounds has proved of little use in taxonomy.

As in the case of sterols, some of the early work on sponge pigments cannot be accepted as accurate until repeated by modern methods. However, certain interesting pigment types do occur in sponges and they characterize some groups.

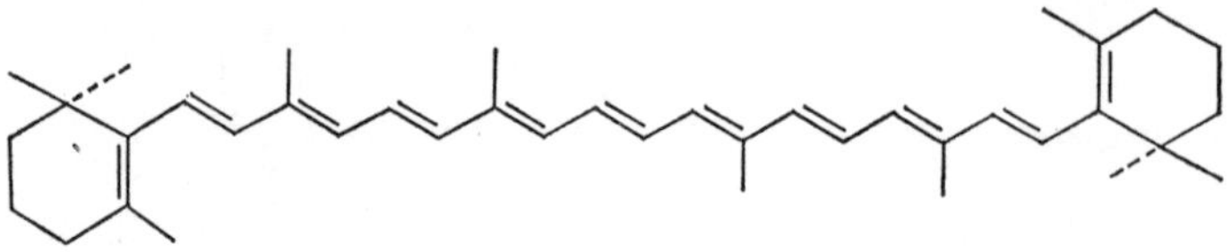

β-Carotene.

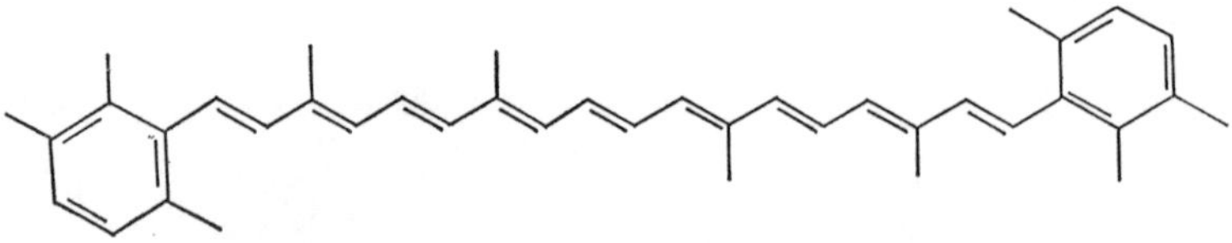

Isorenieratine, an aromatic carotene from *Reniera japonica*.

The commonest sponge pigments are carotenoids, which account for most of the brilliant orange, red, yellow, purple and brown colours. Pigments related to both α- and β-carotene have been isolated frequently. A series of unique carotenoids have been isolated from *Reniera japonica* (Yamaguchi 1957, 1958); these compounds, for example renieratene and isorenieratene, have aromatic rings at each end of the molecule and absorption spectra which are very similar to those of γ- and β-carotene respectively. Two other particular pigments deserve mention. One, a 'uranidine' (Krukenberg 1882) is a yellow pigment found so far only in the Verongiida. In the intact sponge this is a dense sulphur yellow, but immediately the sponge is damaged a navy blue colour is developed which quickly becomes a deep red-purple or black. This sequence is striking and rapid. The chemical basis of this colour change obviously involves oxidation of the pigment, the precise structure of which is so far unknown.

The other pigment of particular interest is the dense royal-purple pigment found in some Hadromerida, specifically species of *Cliona* and *Spirastrella*. This has been reported to be a porphyrin, 'spongioporphyrin' by MacMunn (1900), but its precise nature is not known. It is so clearly the same pigment in these two genera that its occurrence serves further to stress the relatedness of the two families in which the ability to excavate calcareous material is best developed.

When preparing sponge extracts for free amino acid analysis, Bergquist and Hogg (1969) and Bergquist and Hartman (1969) noted the occurrence of several water-soluble pigments. These are not uncommon in the Halichondria and Tethyidae, but so far their composition is unknown. It is possible that they are carotenoids complexed with protein and thus rendered water soluble.

8 The fossil history and phylogenetic relationships of sponges

8.1 Introduction

The important events in the evolution of the Porifera took place in Pre-Cambrian time, for in the early to mid Cambrian representatives of all recent groups, and some others which appear not to have survived, are present and already widespread (Finks 1970; Ziegler and Rietschel 1970). The story of sponge evolution insofar as it is documented in the fossil record is thus essentially a Paleozoic story. The structural types which lived until the late Mesozoic have persisted in great part until recent time. Patterns of numerical dominance changed and families became extinct, but the major groups remained and are recognizable today.

Consideration of the evolution of the Porifera can be confused by uncertainty as to the precise definition of a sponge. Even if we ignore minor fossil groups it is still important to decide whether to include the Receptaculitida, Archaeocyatha, Sphinctozoa, and Stromatoporoidea within the Porifera, and where to classify the Heteractinellida (=Heteractinida) within the phylum. Disagreement on the detailed relationships of all of the above groups still dominates the paleontological literature.

The discovery of living Sphinctozoa and Sclerospongiae, and the recognition of the relationship between recent Sclerospongiae and many fossil Stromatoporoidea and Chaetitida, allow a definite statement with regard to the fossil members of these groups. They were sponges.

The Receptaculitida have been studied by Ziegler and Rietschel (1970), who expressed the view that these spherical or conical organisms had no relationship to the Porifera and were more likely to have been calcareous algae. Sponge affinities for the group had been urged on the basis that the wall of *Receptaculites*, the best-known genus, was built up of particulate elements termed 'meromes', surrounding an inner cavity filled either with sediment or calcite. The shape of the meromes is reminiscent of sponge spicules. However, the shape has been shown to be caused by artefact of preservation. The meromes were probably complex organs which had a distinct growth pattern and calcareous composition during the life of the organism. Receptaculitida, therefore, possessing neither pores nor a canal system and having no spicules, cannot be regarded as sponges.

The position of the Archaeocyatha is again difficult to resolve in the face of severe differences of opinion among paleontologists (Ziegler and Rietschel 1970;

Zhuravleva 1970). The former authors argue for the inclusion of the group within the class Calcarea, at least until more data become available to clarify the position. Zhuravleva on the other hand separates the Archaeocyatha widely into a separate subkingdom, the Archaeozoa. Archaeocyathan fossils were conical or cylindrical cups with two perforated walls, inner and outer, which enclosed a space, the intervallum. There was a hollow central cavity. The intervallum housed a skeleton of variable form; rods, radial porous struts termed parieties, and tabulae (Fig. 8.1). No spicules were present and the entire skeleton was composed of microgranular calcite.

The pores in Porifera and Archaeocyatha appear to be of very different nature. In sponges, pores arise in soft tissue around which initially discrete skeletal elements are disposed; in Archaeocyatha, it is apparent from excellent fossils of young stages that the pores are present in the initial stages of skeleton development as basic elements in the skeletal layers. This, in view of the lack of spicules and the sophisticated compartmentalization of the body, suggests that Archaeocyatha were not Porifera. Zhuravleva's view that they are a separate evolutionary development seems logical, for in different structural features the group expresses

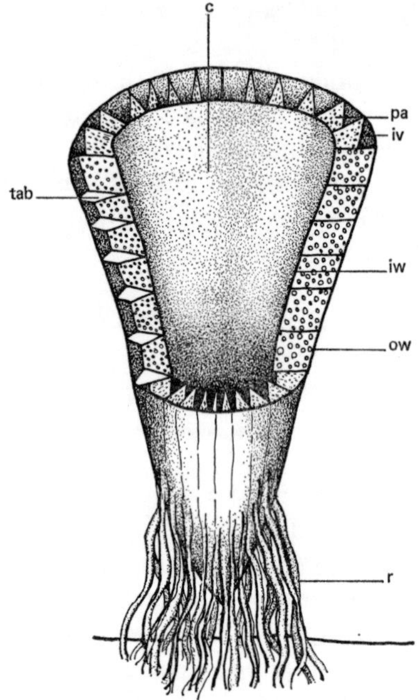

Fig. 8.1 Schematic diagram of a typical archaeocyathan. Vertical section partly cut away to show the structure between the inner and outer walls.

ow, outer wall; iw, inner wall; iv, intervallum; c, central cavity; pa, pariety; tab, tabula; r, rooting processes.

(Redrawn from Moret 1952.)

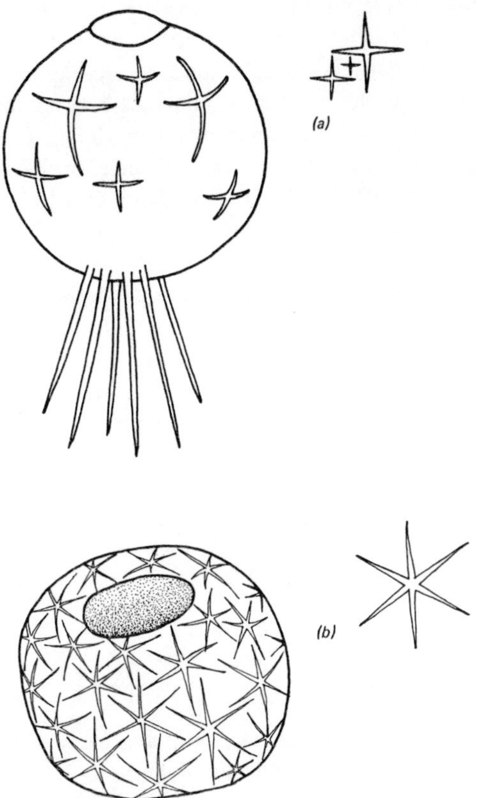

Fig. 8.2
(*a*) *Protospongia*, a genus which represents the most primitive type of hexactinellid construction. It was a thin-walled sac supported by siliceous spicules which were quite discrete. All spicule rays were in one plane which was always paratangential to the body surface.
(*b*) *Eiffelia*, one of the best-preserved genera of Cambrian Herteractinellida. The sponge was globular and thin-walled with octactinal spicules in which two rays were represented only by nodular lumps, if at all. This gave the spicules an overall hexactinal appearance.
(Redrawn from Finks 1970.)

convergence with almost all groups of the lower Invertebrata which were known in the lower to mid Cambrian (Protozoa, Sphinctozoa, Hexactinellida, Stromatoporoidea and the rugose and tabulate corals). She argues that the Archaeocyatha were one of the earliest distinct morphological types to become established. It is not necessary here to express a view on the status of the group at the subkingdom level, but their affinities do not seem to be with the Porifera. The Archaeocyatha were extinct by the mid Cambrian.

Most paleontologists (De Laubenfels 1955; Finks 1970) concur in viewing the Heteractinellida as a relatively short-lived group of sponges which had spicules of a unique type. Some workers consider the Heteractinellida to be related to the Hexactinellida, and stress the external similarities between the heteract *Eiffelia* (Fig. 8.2*b*) and the hexactinellid *Protospongia* (Fig. 8.2*a*). It is difficult to reconcile this view with that of most other workers (Ziegler and Rietschel 1970; Rigby and Nitecki 1975; Rigby 1976). These authors place all Heteractinellida in the class Calcarea, but divide the genera differently into families and interpret their relationships differently. The spicules in early forms were hollow and the rays were articulated to a centrum. This articulated junction remains evident as a suture line similar to that seen in some modern Calcarea. Also, more advanced later heteractinellid fossils were definitely calcareous and these seem clearly to be derivatives of the debated Cambrian forms. These arguments are accepted and the Heteractinellida are treated as primitive Calcarea.

The Heteractinellida were extinct before the end of the Palaeozoic and contribute little to an understanding of the relationship of ancient to recent sponges, other than to demonstrate that Calcarea with free spicule skeletons were present in the lower Cambrian.

8.2 Evolution of the major groups of Porifera

8.2.1 Hexactinellida

The first recognizable Hexactinellida occur in the early Cambrian. They were all of the *Protospongia* type, extremely thin walled sac-like sponges in which the clearly separate spicules formed a single layer (Fig. 8.2*a*). This dispersed skeleton could not support a thick body wall and it has been suggested that the choanocyte layer probably had an asconoid arrangement. The spicules were stauractines, tetractinal forms in which all four axes lay in the same plane. In modern hexacts stauractines are often huge, they form the wide-meshed framework of the skeleton in *Euplectella* (Fig. 1.5*d*, p. 24), for example. These spicules are interpreted as being derived from hexactinal forms by the suppression of one axis. It is possible, however, that stauractines were more primitive than true hexactinal spicules in the Hexactinellida, since they are the only megascleres to occur in larvae of living hexacts (Fig. 4.7, p. 113). Hexactinal spicules are found in the early Cambrian and the fact that they are present implies that sponges with a thicker body wall structure had evolved at that time, but the thin *Protospongia* type was certainly more common. In the upper Cambrian a fossil with a coherent skeleton of hexactinal spicules is found in *Multivasculatus*, and the advent of advanced, quite modern hexactinellid construction is seen in the mid Ordovician *Brachiospongia*. This genus had large, specialized, hypodermal spicules covered by smaller autodermal types (Fig. 5.1*f*, p. 141) and a body wall substantially thicker than any known Cambrian form. From this assemblage, modern lyssacines such as *Euplectella* probably derived.

In the Ordovician a second hexactinellid lineage which derives clearly from *Protospongia* makes its appearance. These sponges retained the uniform stauract spiculation with parallel orientation. In the mid-Palaeozoic the dominant group in this lineage was the Dictyospongiidae in which the stauracts were lengthened greatly and more closely opposed to one another. An inner layer of longitudinal and horizontal bundles of monaxon spicules was sometimes present, but the sponges remained thin walled. In the Permian two divergent lines had developed from this stock, both achieving a stronger skeleton by emphasizing the monaxons on the one hand or the stauracts on the other. One group, of which *Stereodictyum* is typical, had a rigid, thick wall in which the monaxons were multiplied enormously and the stauracts eliminated. The other group, represented by *Microstaura*, enlarged the outer layer of parallel stauracts into a regular cubic mesh of hexacts. The spicules in this form were still separate, but were closely packed and overlapping. There is very little difference between this type of body wall and the continuously fused cubic mesh of the true dictyonine hexactinellids such as *Farrea* which appeared in the Triassic and which persist to the present.

Thus, by the early Mesozoic, Hexactinellida which equate in megasclere spiculation and skeleton organization with modern forms of both dictyonine and lyssacine type had become established.

A comment should be added on the Paleozoic history of hexactinellid microscleres since the classification of recent forms is based on the mutual exclusion of hexasters (Fig. 1.5*f*, p. 24) and amphidiscs (Fig. 1.5*b*). Some of the Carboniferous Dictyospongiidae had, in addition to hexasters, paraclavules which are shaped like one-ended amphidiscs. Unequal-ended amphidiscs, termed hemidiscs, and true amphidiscs recorded from the Carboniferous, are almost as ancient as hexasters. The occurrence of paraclavule and hexaster and of amphidisc and hemidisc together in some fossils suggests that possibly the modern Hexasterophora and Amphidiscophora could have diverged from a common stock during the Paleozoic, and that they could be the sole survivors of a number of early lineages in some of which hexasters and amphidiscs could have occurred together. There is no fossil evidence yet that this was the case.

Two clear lines of evolution within the dictyonine Hexactinellida are represented by the Lychniscosa, where the huge spicules making up the fused skeleton developed buttresses at the point of intersection, and the Hexactinosa where this complex nodal structure is lacking. The former occurred first in the Triassic and reached a peak in the Cretaceous. They are represented today by a single family, the Aulocystidae. They originated either from early Hexactinosa or independently from primitive lyssacine stock. The Hexactinosa were rare in the Triassic and reached a peak in the Cretaceous when many modern families were established. The group has declined somewhat to the present, but since its origin it has been by far the most abundant hexactinellid group and remains so to the present.

Phylogenetic relationships within the Hexactinellida are difficult to evaluate. Clearly the lyssacine skeleton with separate spicules in thin-walled sponges is primitive, but tendencies to close packing and fusion of the spicules and thicker-walled construction can be traced in several stocks. The phylogenetic significance of microsclere type is probably almost negligible, since both amphidisc and hexaster were present in Paleozoic time, where they occur in related fossils. Present classification is one of convenience, and further biological information is required before real relationships between living hexact groups will become apparent.

During their long history the Hexactinellida have migrated from littoral waters, where they were common in the Paleozoic, to the deep oceans where today, particularly in the Antarctic, they are major components of the benthos. It is tempting to associate the early stages of migration with the evolution of thicker-walled construction during the Cambrian. Early forms such as *Protospongia*, with thin walls and diffuse spicule nets, are found best-preserved in black shale, a quiet-water deposit. Evolution of thicker body walls would permit the development of a more efficient canal system and hence more efficient feeding. Also, since such sponges would be more robust they could move into environments of high energy. This was achieved by the late Cambrian and lower Ordovician, at which time Hexactinellida and some other sponge groups colonized shell banks and reefs as well as quiet waters.

8.2.2 Calcarea

An excellent interpretation of the divergent evolutionary trends within the Calcarea is given by Ziegler and Rietschel (1970) (Fig. 8.3). In their view the Heteractinellida are to be included within the Calcarea, and in this way it is established that Calcarea, were present with all other sponge classes, in the lower to mid Cambrian. Three main structural types of Calcarea can be distinguished: those emphasizing discrete spicules, those which developed calcareous fibres and those which evolved calcareous walls. Calcarea which developed a calcareous skeleton made up of discrete spicules are first identifiable in the lower Cambrian

Fig. 8.3 Representation of the major evolutionary trends in the class Calcarea. The evolution of skeletons based on free spicules characterizes the Heteractinellida (Chancelloridae and Octactinellida) and recent Calcinea and Calcaronea (Dialytina for Ziegler and Rietschel).

Organization which adds calcareous fibres to spicule skeletons is evident in fossil Pharetronida (Inozoa) and in some recent forms.

Development of calcareous walls which subdivide the living tissue into units characterizes the Sphinctozoa.

The geological periods during which each group is represented in the fossil record are shown at the left. All groups which survived the Paleozoic have persisted to the present.

C = Cambrian, O = Ordovician, S = Siluvian, D = Devonian, Ca = Carboniferous, P = Permian, T = Triassic, J = Jurassic, Cr = Cretaceous, Ter = Tertiary.
(Modified from Ziegler and Rietschel 1970.)

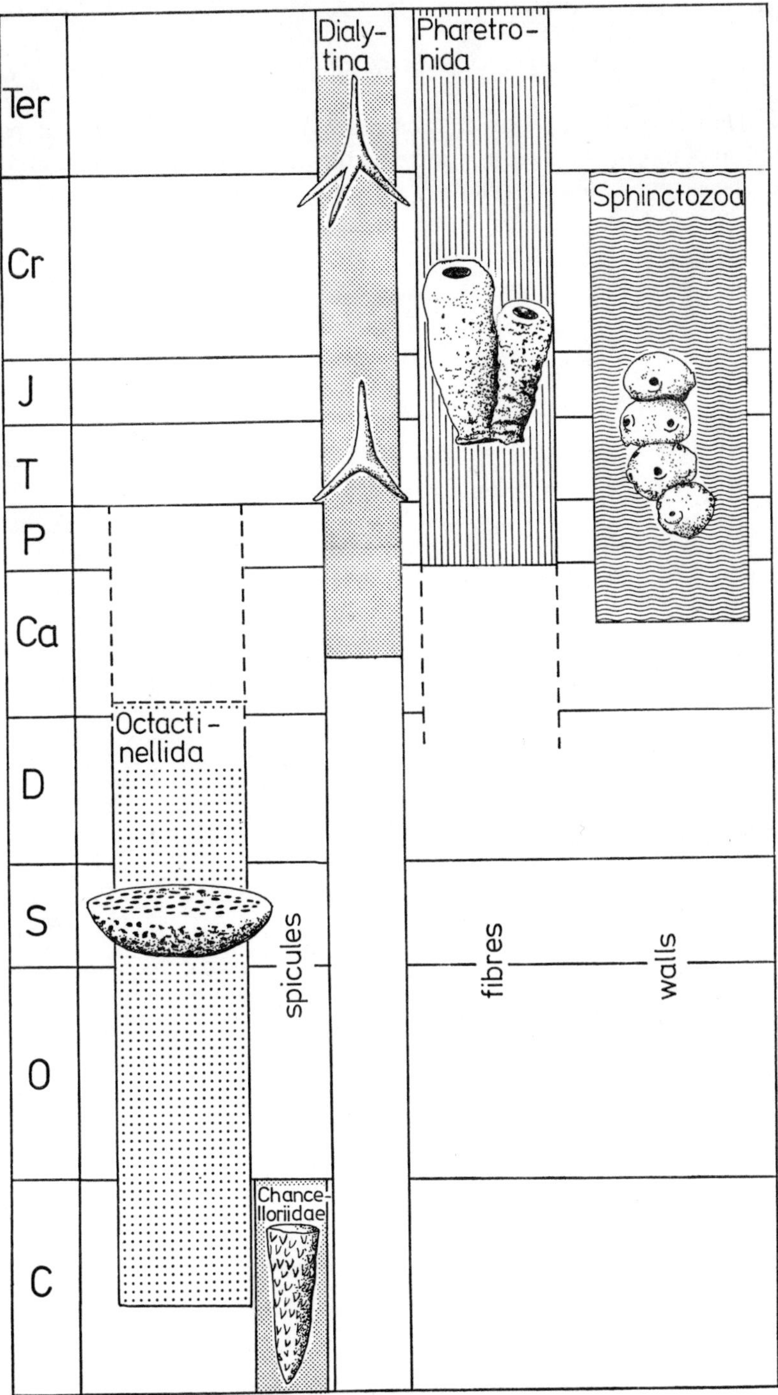

Ter
Cr
J
T
P
Ca
D
S
O
C
Dialy-
tina
Pharetro-
nida
Sphinctozoa
Octacti-
nellida
Chance-
lloriidae
spicules
fibres
walls

Plate 11

(a) *Astrosclera willeyana* (Sclerospongiae).

(b) Details of the surface of *Astrosciera* to show the astrorhizal patterns imprinted into the calcareous matrix by the exhalant canals and oscules of the superficial sponge tissues. (Photos G. Batt.)

a

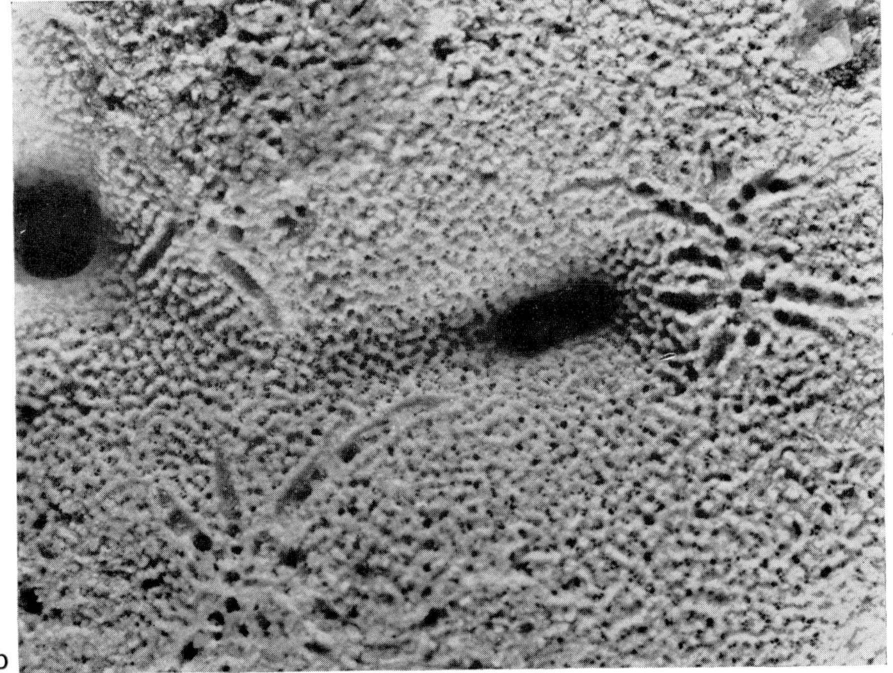

b

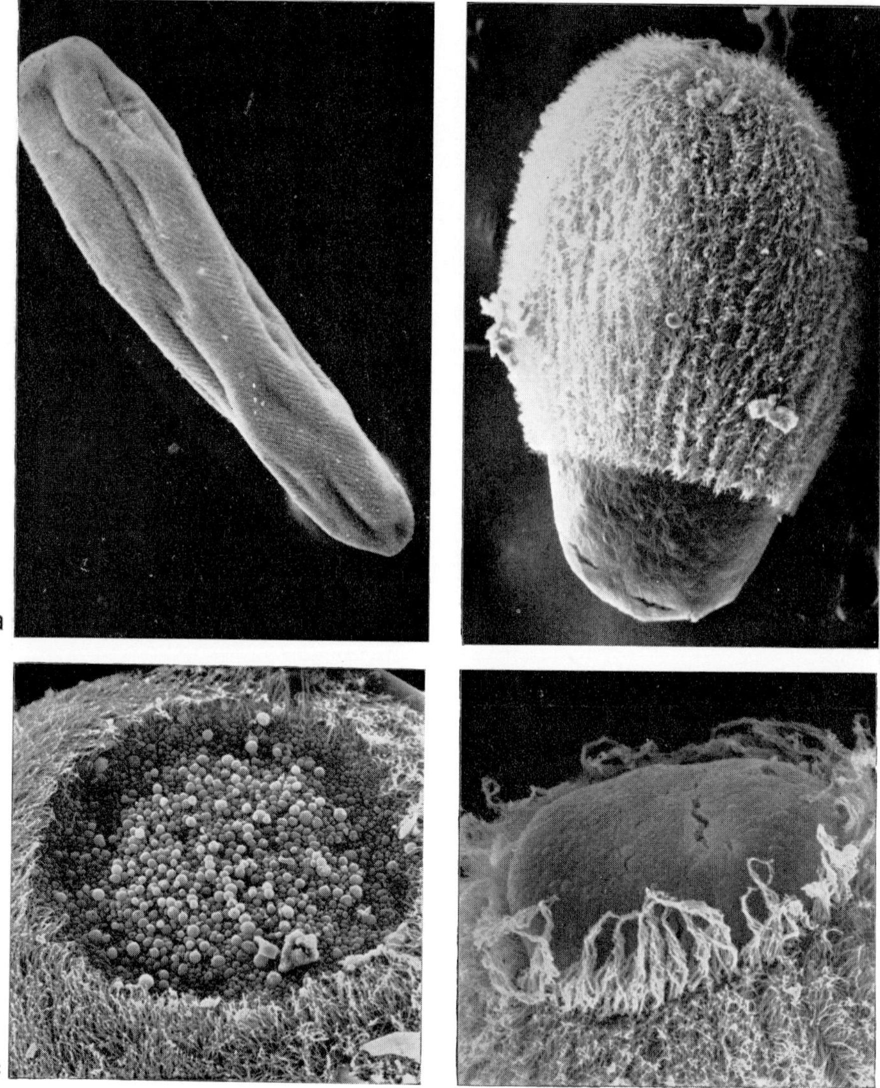
a
b
c
d

Plate 12
Scanning electron micrographs of larvae of some Demospongiae.

(a) *Halichondria moorei* (Halichondrida). The larva is entirely ciliated. Metachronal ciliary waves can be seen passing obliquely round the body of the larva.

(b) *Lissodendoryx isodictyalis* (Poecilosclerida). The larva has a large posterior area bare of ciliar.

(c) *Aplysilla rosea* (Dendroceratida). The anterior pole of the larva just prior to settlement. The unciliated area exudes secretary material which could have a role in initial attachment.

(d) *Haliclona* sp. (Haplosclerida). The posterior pole of the larva showing the small bare area and fringe of long cilia.

Chancelloria which had large calcareous spicules constructed of up to nine hollow rays, each of which articulated to the spicule centrum. The junction between ray and centrum was marked by a noticeable suture at the point of intersection. The Chancelloriidae were obviously a specialized offshoot from some simpler but unknown ancestors, and were most abundant in the Cambrian. Until recently they were considered to be extinct by the upper Cambrian, but Rigby and Nitecki (1975) have described a mid Carboniferous sponge *Zangerlispongia* which, in its thin-walled conical construction and two-dimensional six-rayed spiculation, is very like *Eiffelia*, which the same authors place in the Chancelloriidae. *Zangerlispongia* did not have sutured spicules and in this case it is argued that the thickening of spicule elements in this genus obscures the primitive condition in which thin-walled hollow spicule elements are sutured.

In the mid Cambrian, the Octactinellida appeared. These are common and widespread fossils in Paleozoic reef and shelf sediments and are typified for Ziegler and Rietschel (1970) by the early Cambrian *Eiffelia* (Fig. 8.2*b*). This was a vase-like sponge with a skeleton of interlocking octactinal calcareous spicules, in which two rays were reduced to nodes. In Devonian genera, monaxonid spicules had been added to the octactines, and in the Carboniferous family Wewokellidae calcareous triactines and tetractines had replaced the octactines. The latter group with their triact spicules and pronounced dermal layers reinforced by triacts can be strikingly similar to some Mesozoic Pharetronida (Inozoa). It is possible that the latter group was derived from the Wewokellidae during the Carboniferous.

The classification of *Eiffelia* in the Octactinellida by Ziegler and Rietschel (1970), and in the Chancelloriidae by Rigby and Nitecki (1975), is not a serious divergence of opinion. The effect of the later authors' discovery of *Zangerlispongia* has been to bring this whole heteractinellid group closer together, while previously the lower Cambrian forms could only be related doubtfully to later fossil types. The Heteractinellida were extinct by the lower Permian.

The first unequivocal fossil occurrence of a member of the Calcaronea is the record of *Leucandra* from the lower Jurassic. This sponge exhibits effectively modern construction and it is assigned to a recent genus. Previous records of detached spicules (Carboniferous, Triassic) are, however, very dubious. The Calcinea are not certainly known as fossils.

While Calcarea with discrete spicule skeletons are only rarely fossilized, those members of the group which evolved massive skeletons are very well represented. These are the sponges which persist today as a relict group, the Pharetronida. Within this group the fossil members of the order Inozoa are characterized by the formation of a reticulation of calcareous fibres in which spicules may or may not be present, and members of the order Sphinctozoa have massive calcareous skeletons which always exhibited some compartmentation.

The Sphinctozoa, characterized by the possession of solid calcareous walls subdivided into compartments, first appeared in the lower Carboniferous. At their appearance they were already a diverse group in which three major and quite separate structural types were recognizable. These can all be traced through continuously developing separate lineages into the Permian and thus appear to have been quite separate evolutionary lines. In one family, the Sphaerocoelidae, the calcareous walls contained three- and four-rayed spicules while other groups lacked spicules in the fossil state. This does not prove that they were absent in life when they may have occurred as free elements in the soft tissue which, in living sphinctozoans, extends over the external apical area and fills the interior of the individual segments. All sphinctozoans had pores or larger spout-like openings perforating the calcareous walls and, with their system of pores and channels, appear to have functioned much like modern Calcarea (Fig. 5.7, p. 153).

The Sphinctozoa extended from the Carboniferous to the Cretaceous as fossils. They were common in Permian and Triassic time and declined throughout the Jurassic and Cretaceous. As yet we do not know whether the living genus can be allied with any known fossil group or whether it represents a new structural type. In the Permian, sphinctozoans were reef-builders, and all fossil forms appear to have lived in shallow-water, shelly habitats. Significantly, living Sphinctozoa were discovered in shallow caves on coral reefs.

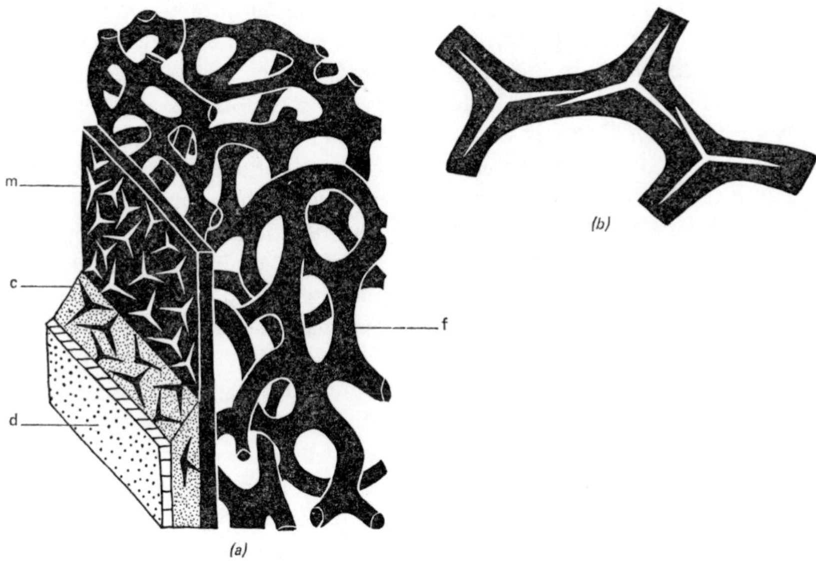

Fig. 8.4 The organization of Pharetronida as interpreted from fossil material.
(*a*) A diagrammatic representation of a Jurassic pharetronid. There is an outer dermal membrane (d), and a spicule-containing cortex (c), which is separated from the central network of calcareous fibres (f) by a calcareous membrane which contains spicules (m).
(*b*) Enlarged view of a fibre to show the enclosed triradiate spicules.
(Redrawn from Ziegler and Rietschel 1970.)

The third type of organization seen in the evolution of the Calcarea has calcareous fibres in addition to a massive secreted skeleton and free spicules (Fig. 8.4). These sponges, the Pharetronida (Inozoa), appeared first in the Permian and were abundant in the Mesozoic. Most genera died out at the end of the Cretaceous, but about ten genera have persisted until recent time. Like the Sphinctozoa, the Inozoa were Permian reef-builders which in modern times have retreated to become cave-dwellers in coral biotopes. On the basis of the discovery and description of several recent genera of Pharetronida Inozoa (Vacelet 1964, 1967), it has proved possible to assess the morphological significance of the calcareous spicule-containing fibres. Although used by paleontologists to diagnose the Inozoa, these structures have been the subject of much discussion centring on whether they were actual structures during life or whether they are artefacts of preservation. Some recent species belonging to the family Minchinellidae have spicules which are welded into a mass by the secondary deposition of a calcareous cement. These correspond precisely with the 'fibres' seen in those fossil forms which can be assigned to this family. It is significant in this group that only a single type of spicule is incorporated into the 'fibres'. However, in other groups such as the Murrayonidae, the fibres of the living species lack spicules while those of the fossil representatives contain spicules. This seems certain to be an artefact. The fibres in the fossil species probably represent casts of the lacunae between the elements of the primary, aspiculous, calcareous network. These spaces in life contain the soft tissues and free spicules of the sponge and after death would be infilled by sediment. In this way it is possible to envisage how fossil sponges might appear to have possessed 'fibres' in which spicules of several different types occurred. Lastly, there are recent forms belonging to the family Lelapiidae where there are no fibres, only simple unfused spicules in tracts. These are fragile sponges which do not occur as fossils. The presence of 'fibres' which contain spicules therefore certainly cannot characterize the living Inozoa. This serves to illustrate the great difficulties encountered in integrating a classification based on fossilized material with one based on living material.

The fossil record offers little that is unequivocal to an interpretation of the inter-relationships of the three subclasses of the Calcarea. It is conceivable that ancestors of the modern Calcinea and Calcaronea were more ancient than those of the Pharetronida (Inozoa and Sphinctozoa), but such sponges would only fossilize under exceptional circumstances. A relationship between the Chancelloriidae, Octactinellida and modern Calcinea and Calcaronea cannot be urged seriously, but it is possible that some Carboniferous Octactinellida were ancestral to the inozoan Pharetronida. However, if the distribution and abundance of the various subclasses of the living Calcarea are considered, in conjunction with their morphology, histology and development, it is more logical to regard the inozoan Pharetronida as ancestral to the Calcinea and Calcaronea. Not only are the Inozoa at present a relict group flourishing only in dark caves, but they also exhibit a mixture of the histological and larval characteristics which can be used to define the other two subclasses (refer Chapter 5). This distribution of characters lends itself to the idea that the Calcinea and the Calcaronea arose from different stocks within the Inozoa. The Sphinctozoa could prove ultimately to be a fourth subclass when histological and larval characters of the living species are available for detailed comparison with those of other Calcarea.

8.2.3 Sclerospongiae

The rediscovery of *Ceratoporella, Merlia,* and a series of previously undescribed but related genera in cryptic and deep-reef environments in Jamaica (Hartman and Goreau 1970), led to the establishment of the class Sclerospongiae. At the time of the original proposal, all known genera could be accommodated within one family, the Ceratoporellidae. Subsequently a second order, the Tabulospongida, was set up to include *Acanthochaetetes wellsi,* a living species of a fossil genus in which the tabulate calcareous skeleton is calcitic rather than aragonitic (Hartman and Goreau 1975). Hartman and Goreau (1970), when proposing the new class, drew attention to the strong morphological resemblances between the recent Sclerospongiae and the fossil Stromatoporoidea, a group which lived for 400 million years from the mid Cambrian to the end of the Mesozoic. During this period they were reef-builders inhabiting shallow waters near open ocean shorelines.

The affinities of the Stromatoporoidea have always posed a problem for paleontologists mainly because the hard fossilized skeleton provides few substantive indications as to the organization of the soft tissue in the living animal. Although the group has been most frequently allied with the Hydrozoa, and is sometimes considered to have evolved into the Milleporina, there is no trace in any known stromatoporoid of the partitioning of the body, or coenosteum, as it is termed, into the individual units which of course characterize coelenterate organization (Stearn 1972). There is no evidence in the structure of any fossil to support the idea that Stromatoporoidea actively caught food, as do corals, rather than filtered it, as do the sponges. Many stromatoporoids have tabulae and dissepiments as skeletal partitioning structures, and these are certainly coelenterate characteristics. However, some Sclerospongiae (*Acanthochaetetes* and *Merlia*) exhibit tabular structure and therefore it is possible to regard this characteristic as an ancient one, retained in certain sponge groups, but emphasized only in coelenterate stock. Many paleontologists have indeed suggested that stromatoporoids were sponges rather than coelenterates.

While Sclerospongiae resemble modern Demospongiae in their histology and general spicule shape, and in possessing spongin fibre, it has not proved possible to point to any one group within the Demospongiae to which they might be related. The spiculation of sclerosponges is in some cases unique, and in any case so diverse, that to ally them with Demospongiae we would have to envisage the parallel development of a solid calcareous skeleton in several lines of Demospongiae. There are substantial differences in organization between the two groups. The Sclerospongiae possess a solid calcareous skeleton and have soft tissue organized into units housed in calcareous calicles and served by one or two ostia. These structures do not occur in Demospongiae. Therefore it is reasonable to view the Sclerospongiae as a group which diverged from the Demospongiae at some remote time in the past and which has had a long independent history. The existing species are then interpreted to be the remnants of a group which was once more abundant. The problem has been to suggest which group in the fossil record could be ancestral to modern Sclerospongiae. The Stromatoporoidea is certainly one possible group; the Chaetetida, a group of tabulate corals which lived from the Ordovician to the Cretaceous, is another. Whether the Sclerospongiae have a recognizable fossil history or not hinges substantially on whether these suggested relationships are accurate.

The relationship of sclerosponges to chaetetids seems very well founded (Hartman and Goreau 1975). The two groups have the following features in common: identical size and arrangement of the living tissue units which determine the size and arrangement of the calicles, an identical method of asexual reproduction by longitudinal fission during growth of the calcareous skeleton, and the sharing of the walls between adjacent living tissue units, rather than the condition in which each unit secretes its own calcareous covering. The fact that no spicules have been found in fossil Chaetetida is not a problem in view of the fact that spicules which become entrapped by the growth of the calcareous skeleton in living individuals of *Ceratoporella* are known to erode rapidly, and thus they would not be expected to persist in fossils. Also in *Merlia*, a sclerosponge with a tabulate structure (Fig. 5.27, p. 180), the siliceous spicules always remain above the calcareous base and never become trapped. They are lost when the soft tissue disintegrates. The absence of superficial astrorhizal patterns in chaetetids could reflect the fact that the layer of living tissue above the calcareous skeleton was sufficiently thick to allow confluence of the exhalant canals toward the oscules without exerting pressure sufficient to inhibit upward growth of the underlying calcareous material. Observations of the growth of living Ceratoporellida have shown that astrorhizal patterns in the calcareous material are produced as a result of pressure from the exhalant canals. It thus seems probable that chaetetids and the Ceratoporellidae were related and that the fossil history of these particular sclerosponges extends back to the Ordovician.

Relationship of Sclerospongiae to Stromatoporoidea is suggested most strongly on the basis of comparison with *Astrosclera* (Pl. 11) where the superficial astrorhizal patterns impressed by the exhalant canals are evident, and the detailed microstructure of the calcareous skeleton is identical to that of certain stromatoporoids. Also, *Astrosclera* shows the superimposition of successive astrorhizae during growth of the sponge. This produces exhalant channels which extend deep into the calcareous matrix, a condition which is seen often in fossil forms. If the affinity of Sclerospongiae with certain Stromatoporoidea is accepted it extends their fossil history back to the mid Cambrian. Stromatoporoids were rare in the Ordovician, but spread all over the globe during the Silurian and Devonian when they were a major element in the world invertebrate fauna. After a decline in the upper Paleozoic, the Mesozoic saw a proliferation of the group with considerable changes in species composition.

It still remains for biologists and paleontologists to determine precisely which stromatoporoids and chaetetids were ancestors of the recent Sclerospongiae for it is certain that both groups at present contain some genera which belong in the Porifera and some that belong in the Coelenterata. The discovery of living *Acanthochaetetes* has also made it necessary to scrutinize the composition of the Favositida, an order usually considered to belong within the tabulate coral group. The structure of *Acanthochaetetes* suggests that it is extremely likely that some favositids were sponges belonging to the class Sclerospongiae.

8.2.4 Demospongiae

The earliest Demospongiae from the mid Cambrian were tubular sponges with oxeote spicules arranged either as an isodictyal mesh or in parallel and anastomosing tracts. Skeletal organization such as that seen in *Hazelia* (Fig. 8.5*a*) would certainly require spongin to bind the spicule skeleton. The Demospongiae were the first sponge group to undergo a major adaptive radiation, at least part of which involved the development of lithistid skeletons which, because of their compact structure, remain as fossils. Lithistid demosponges appear in the early Ordovician, at which time two principal stocks were established: the Tetracladina and the Eutaxicladina.

The Eutaxicladina in the Ordovician had already diversified into two major groups, those related to *Astylospongia* and those related to *Hindia*. *Hindia* was built of concentric shells of spicules, each of which had three expanded feet abutting closely on each other (Fig. 8.5*c*). In addition, radial and tangential oxeas were present and these could have supported a superficial dermal skeleton. *Astylospongia* had six-footed spicules (Fig. 8.5*b*). Both *Hindia* and *Astylospongia* develop toward Mesozoic and later forms which possessed tetraxonid dermal spicules and asterose microscleres.

The Tetracladina throughout their long history showed a persistent trend toward decreasing regularity of the spicule net, associated with a change in the principal type of spicule present. In the Ordovician, this group is represented only by types like *Archaeoscyphia* (Fig. 8.5*d*), in which the very beginning of a lithistid skeleton is detectable. The skeleton remained an isodictyal mesh as in *Hazelia*, but the ends of the monaxonid spicules had developed terminal articulations. Later Silurian forms had a less organized skeleton, but had developed true tetraclone or four-branched spicules: a progression can be traced through to the Permian types in which all trace of a regular skeleton had gone and completely modern tetracladine or tetracrepid spicules were present. Another line, which had its origin in the Carboniferous and Permian, and which can be traced through to modern forms, is the Rhizomorina typified by *Haplistion* (Fig. 8.5*e*). These are lithistids in which the body plan is regular and radial and the rhizoclone spicules, or monocrepid roughened desmas, are organized into uniform tracts. Mesozoic forms tend toward a diffuse skeleton in which there

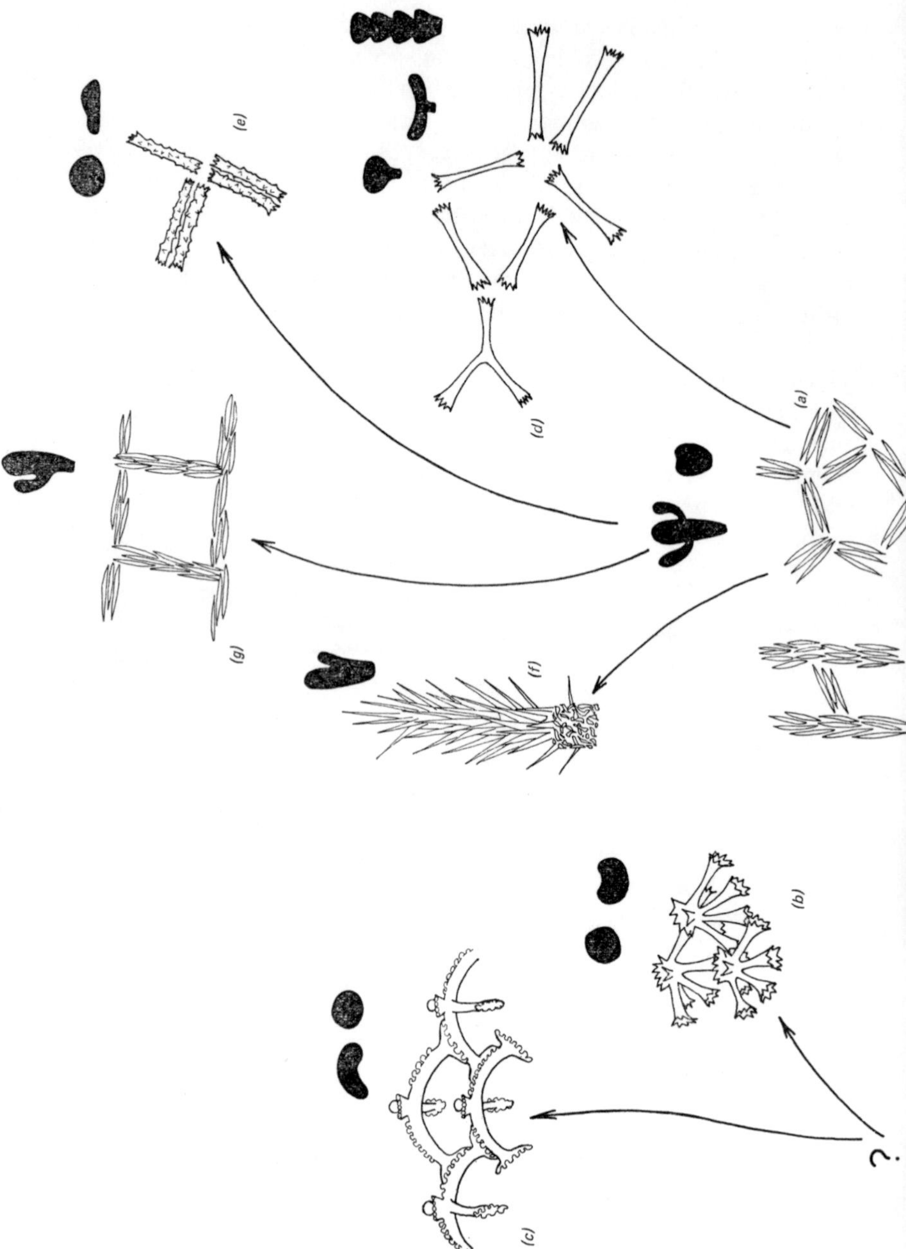

Fig. 8.5 The principal types of skeletal arrangement found in Paleozoic Demospongiae. The form of the whole sponge is indicated in silhouette and the postulated relationships between the groups indicated by arrows.
(*a*) *Hazelia*; (*b*) *Astylospongia*; (*c*) *Hindia*; (*d*) *Arehaeoscyphia*; (*e*) *Haplistion*; (*f*) *Saccospongia*; (*g*) *Heliospongia*. (Redrawn from Finks 1970.)

is no radial arrangement, and towards the development of dermal spicules and distinct inhalant and exhalant surfaces.

Primitive tetracladine Lithistida were early- to mid-Ordovician reef-builders as well as being common in shallow water shell facies. The lithistid organization in which the siliceous spicules interlock clearly proved adaptive in any environment which was exposed to strong mechanical disruption. The first radiation of the Lithistida was completed in the late Ordovician, beyond that time they were displaced by the Stromatoporoidea and were never reef-builders again.

One other important sponge lineage which appeared in the middle Ordovician was represented by *Saccospongia* (Figs. 8.5*f*, 8.6). This sponge had an axial spiculation of styles arranged in plumose fashion in spicule tracts which formed a reticulate net. The axis was surrounded by a layer of simple desmas. This is

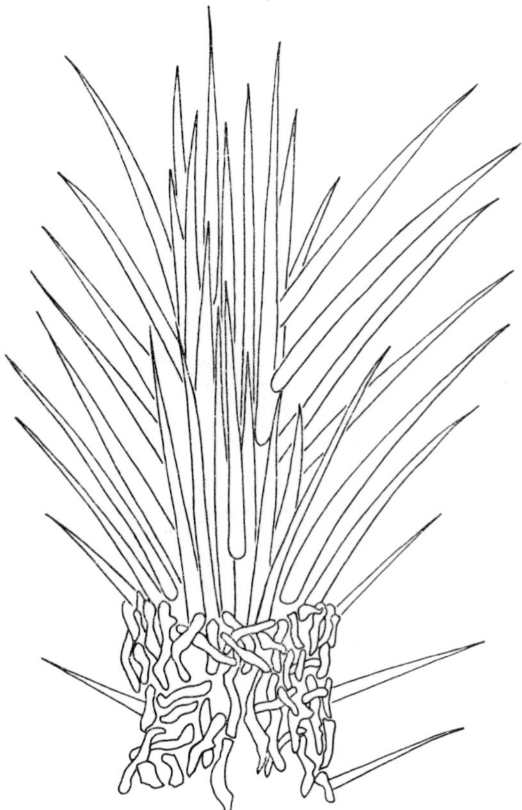

Fig. 8.6 Diagrammatic representation of a spicule tract of *Saccospongia laxata* showing the relationship between the primary axial styles arranged in plumose fashion, the coating of desmas around the axis of the tract and the plumose, extra-axial styles. (Redrawn from Finks 1967.)

the first fossil record of the plumose skeletal architecture which is similar to that preserved today in the Axinellida and in some Poecilosclerida (Mycalidae). Some modern genera such as *Monanthus* and *Petromica* are very similar to the Ordovician *Saccospongia.*

It is not unreasonable on the basis of the fossil types described above to interpret several of the major lines of modern Demospongiae as being already established in the lower Paleozoic. *Hindia* and its relatives foreshadow the Tetractinomorpha, and *Saccospongia* the Axinellida, while *Hazelia* and the later *Heliospongia* (Fig. 8.5*g*) are prototypes of the Ceractinomorpha. The modern Lithistida clearly derive from many lines (Fig. 8.7). By the Cretaceous, representatives of all orders of modern Demospongiae were established (Rezvoi, Zhuravleva and Koltun 1971).

When considering relationships within the Demospongiae, in the light of their fossil history, the obvious advantage in fossilization conferred on lithistid sponges and on some Choristida must not be overlooked. The solid desma skeleton of the former is ideally suited to survive fossilization and the interlocking superficial triaenes of the latter should, to a lesser degree, serve the same purpose. Consequently the absence of fossils referable to the Choristida before the Carboniferous supports the suggestion made by Finks (1970) that these sponges were a later development from among the Ordovician eutaxicladine Lithistida (Fig. 8.7). The fact that the most ancient, mid Cambrian Demospongiae such as *Hazelia* had an isodictyal skeleton of monaxonid spicules, which in life would require spongin fibre to bind them, argues strongly for the view that sponges which could be ancestral to modern Ceractinomorpha have been separate from other Demospongiae since that time. Regrettably very few fossils are left by

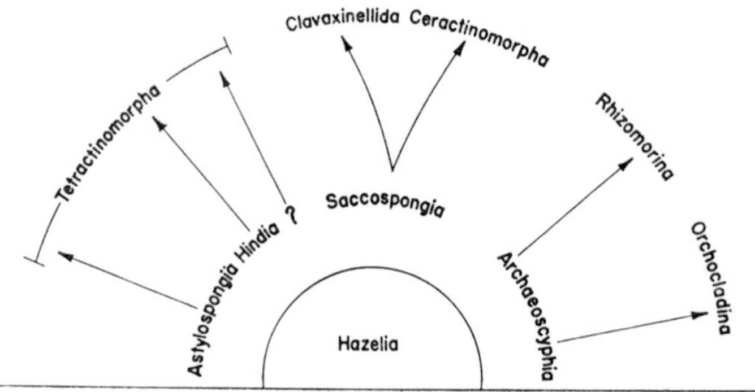

Fig. 8.7 Some postulated lines of descent of living Demospongiae from Cambrian and Ordovician genera. Single genera are used to typify groups of fossil sponges, not to imply that they specifically give rise to the later groups. (From Finks 1970.)

sponges with loose fibre and spicule skeletons. On the other hand, the homogeneity of the recent Ceractinomorpha in terms of reproductive pattern and biochemistry suggests that, with the exception of the Verongida, the diversification of modern groups within the Ceractinomorpha has been a relatively recent event. Some modern families (e.g. Crellidae, Myxillidae, Biemmidae, Haliclonidae, Halichondridae) are recognizable in the Cretaceous, but most do not appear in the fossil record until the Quaternary. The possibility that the development of fibre rather than spicule skeletons has, since the Cambrian, taken place more than once in the evolution of Ceractinomorpha is quite real. It is likely that the Verongida represent a survival of such a long separate line of development.

Because it is possible to argue that sponges with radial architectures and with tetractinal spicule form should fossilize if present, and because they are not found until the Carboniferous, it seems likely that monaxonid not tetraxonid megascleres were the primary skeletal elements in the Demospongiae. Several lines of evolution based on this premise can be suggested, involving the addition of other spicule types to primitive monaxons, which as we have seen are dispersed widely in the Tetractinomorpha as well as characterizing the Ceractinomorpha. The stock which gave rise to the mid Ordovician *Hindia* showed a clear trend toward radial skeletal organization where oxeas were accompanied by multifaceted desmas. Mesozoic lithistids related to this stock had tetraxons and asterose microscleres. Thus the Choristida are likely to have arisen from similar forms which either secondarily lost the desmas or never developed them. The Homoscleromorpha, lacking true monaxonid spicules and having small specialized tetracts are probably quite a separate group. This is also suggested by their possession of peculiar amphiblastula larvae. Similarly, the Spirophorida, with an extremely pronounced radial body plan, predominantly supported by oxeas, but accompanied by several distinct forms of tetractinal megascleres, should be considered as quite an independent development from early monaxonid stock. The Tetillidae are certainly known from the Cretaceous, but one fossil genus *Choia*, from the mid Cambrian, has a spherical body with a radiate skeleton composed of very long oxeote spicules, which extend well beyond the sponge surface. This sponge has been considered to belong to the Hadromerida (De Laubenfels 1955), but it could equally well have been a member of the Spirophorida. If so, the antiquity of the body plan shown by all modern Spirophorida and indeed some Hadromerida would be established.

The Hadromerida again appear to have had a long independent history. Evidence of the activity of the boring sponges (Clionidae) is present in the Ordovician, but the spiculation of the excavators is unknown. However, since the galleries are identical with those bored by recent clionids, it is likely that the hadromerid line, where the monaxons developed into tylostyles, has been separate from other members of the Tetractinomorpha since at least Cambrian time.

In *Saccospongia* from the Ordovician, the axinellid type of construction is evident. These forms also probably represent a separate evolutionary line.

Certainly the modern members of the Axinellida appear, structurally and biochemically, to fall into several very disparate groups. It is almost as if all those sponges with axial skeletal organization, which are now classified as families of the present Axinellida, diverged extremely early from stock with a structure similar to that of *Saccospongia*. Such a long, separate history could account for present difficulties encountered in classification, both internally, within the Axinellida and in relating the order to other Demospongiae. There is no evidence to support a close relationship of the Axinellida and the Hadromerida, as was implied when the super order Clavaxinellida was proposed (Levi 1956). The Hadromerida and Axinellida both lack tetractinal spicules and have monaxons of quite different type. In both cases the diversity of their microscleres is such that only a very distant affiliation with other tetractinomorph stock can be suggested. The Axinellida also preserve some features such as sigmoid microscleres and a spongin skeleton, which have been emphasized to a greater degree in the evolution of Ceractinomorpha.

It is clear from living genera, and from the fossil record, that lithistid organization is polyphyletic, and that it was probably an adaptation favoured by particular ecological situations in many lines of sponges. It is not productive at this time to comment on the affinities of recent forms and the significant fossil genera have been described earlier.

The homogeneity of the viviparous Ceractinomorpha certainly indicates a burst of relatively recent evolution, possibly stimulated by the evolution of viviparity in some pre-Cretaceous oviparous stock. The Verongida could then be interpreted as having diverged from the main ceractinomorph line prior to this time. *Halisarca* with its simple histology, lack of fibre skeleton, curious flagellated chambers, and simple parenchymella larva is probably as close as recent Demospongiae come to the ancestral viviparous ceractinomorph. It is probable, however, that loss of the spongin fibre skeleton was secondary and that the ancestors of *Halisarca* were more like other recent Dendroceratida. These have weak fibre skeletons and simple eurypylous choanocyte chambers, but are complex histologically in both adult and larva. They are best regarded as an independent evolutionary line from the beginning of the ceractinomorph radiation. Within the Dictyoceratida, the Dysideidae most resemble the Dendroceratida, retaining simple histology and eurypylous choanocyte chambers, but having developed the anastomosing fibre skeleton which has allowed the evolution of bodies of great coherence and size in the Spongiidae.

The association of spicule with fibre which has proved so successful in the Haplosclerida and Poecilosclerida is something which has clearly evolved in parallel with the purely fibrous skeletons of the Dictyoceratida and Dendroceratida, since groups showing modern haplosclerid and poecilosclerid organization are diversified in the Cretaceous. It can be argued that the Haplosclerida have some affinities with the Dictyoceratida through forms with very sparse spicule skeletons (like *Callyspongia*) or with no spicule skeleton (like *Dactylia*), but again this is probably an expression of common ancestry where one attribute

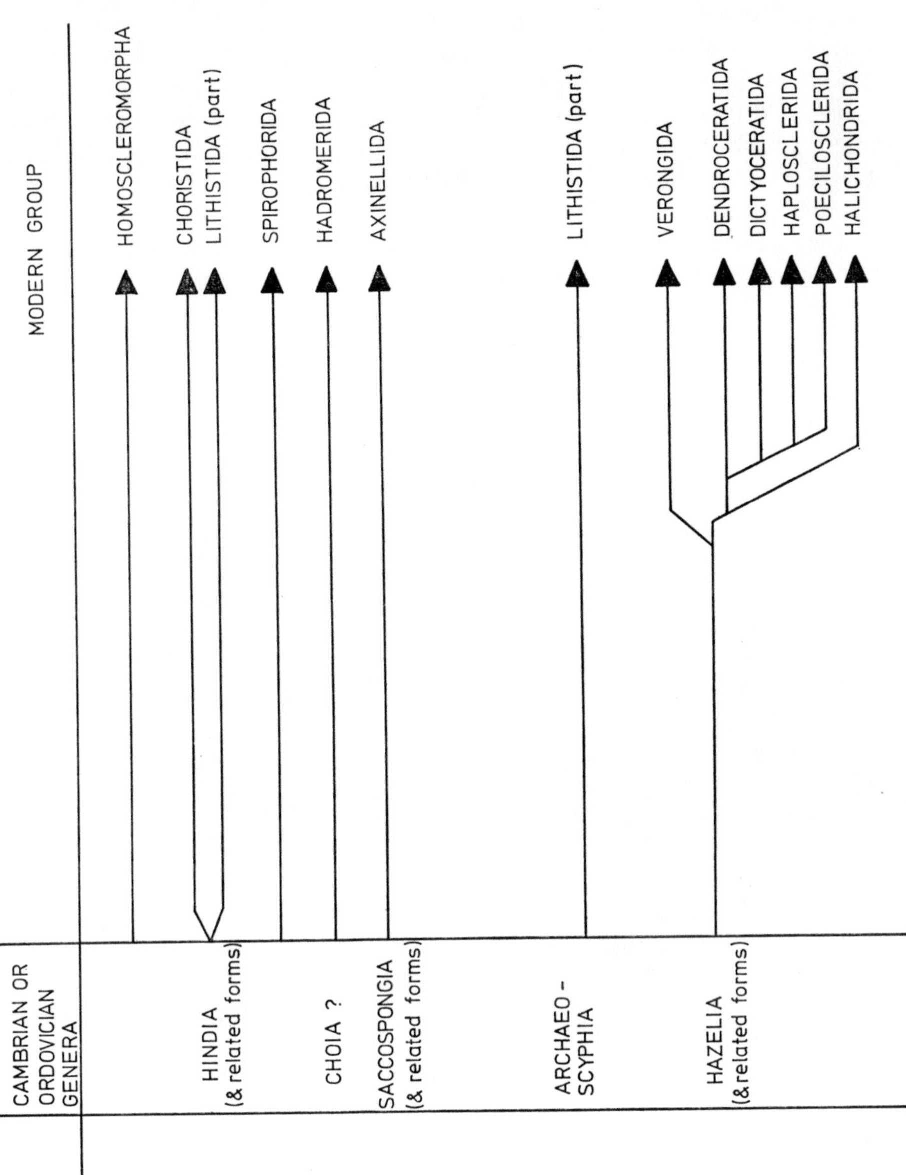

Fig. 8.8 Suggested relationships between early Paleozoic Demospongiae and modern members of the group and an indication of the inter-relationships of modern orders.

is emphasized in one line, and another in a different line. Poecilosclerid sponges have achieved the greatest elaboration of spicule and fibre skeletons to be found within the Demospongiae. The Halichondrida are best interpreted as a group which has never achieved any real elaboration of spongin fibre as a structural element; they remain fragile sponges and are not diverse today except in the intertidal and shallow subtidal areas or in polar regions. It is quite possible that they are polyphyletic. The Halichondridae and the Haliclondiae were separate in the Cretaceous.

The recent fresh-water sponges are grouped into two families; one containing the widespread Spongillidae, the other reserved for the African Potomolepidae (Brien 1969). There is another distinctive taxonomic assemblage of fresh-water sponges recorded only from Lake Baikal, and this is considered by some authors as a separate family, the Lumbomirskiidae. It is clear that there have been at least three separate invasions of fresh water by Demospongiae belonging to the Haplosclerida. It is interesting that the ability to invade fresh water has been confined to one order; the evolution of lithistid skeletons has occurred in several.

In summary, the fossil record lends some support to the interpretation, based largely on the structure and biology of recent Demospongiae, that apart from a group of five orders within the Ceractinomorpha, no one order of Demospongiae can be related unequivocally to another (Fig. 8.8). This has been recognized in the separation of the Homoscleromorpha from the Tetractinomorpha (Levi 1973), but ultimately, other lines at present grouped together, will certainly be separated at the subclass level. A future separation was predicted by Levi (1957) when he discussed the relationships of Demospongiae primarily in the light of reproductive characters. Since that time, a great deal has been discovered about the reproductive patterns of many groups and histological and biochemical comparisons have been made in attempts to interpret relationships. All the evidence is in support of a long independent history of most groups which are now recognized at the ordinal level.

In discussing broad phylogenetic relationships between groups of the Demospongiae, the type of microsclere present in the sponge has been almost ignored. This is quite deliberate. It cannot be stressed too strongly that microsclere spiculation is just one character, it is also a notoriously variable one in living sponges. Classifications based on this feature in the past have proved unsatisfactory. Further, microscleres are relatively rare *in situ* in fossils, and even when free, are only preserved occasionally. Far more insight can be gleaned from studying first the organization of the skeleton, and then its megasclere composition, than by giving primary emphasis to microsclere complement.

8.3 Patterns of numerical dominance

The dominant groups of Paleozoic sponges were the dictyospongid Hexactinellida, orthocladine and eutaxicladine Lithistida and the Heteractinellida. These became extinct by the close of the Permian, a period which saw widespread extinction in marine environments. However, most of the sponges that dominated the Mesozoic had originated during Carboniferous or Permian times. The exceptions to this generalization are the Hexactinosa and Lychniscosa within the Hexactinellida. These groups did not appear until the mid Triassic although hexactinosan structure was foreshadowed in Paleozoic Hexactinellida.

Sponges as a whole achieved their greatest dominance in the Cretaceous. This is not an artefact of preservation for it is reflected in several groups which are readily fossilized because of their coherent skeletons and which are alive today (Finks and Hill 1967). For example, the Pharetronida (Inozoa) had 70 Cretaceous genera and has 10 recent ones; the Lithistida had 150 Cretaceous genera and has 50 recent ones; Hexactinosa and Lychnicosa numbered about 170 Cretaceous genera and there are 120 recent ones. At the same time it is apparent that the fossil record does not tell the whole tale. If we consider the Demospongiae there are around 600 living genera, as opposed to 60 genera of Calcarea and 130 of Hexactinellida; however, only 50 of these living Demospongiae are Lithistida and it is the Lithistida which provide nearly all the fossil record of the class. If the recent proportion of lithistids to other Demospongiae (1:12) prevailed during the Cretaceous, then there were some 1800 demosponge genera of which few, except the 150 lithistids, were preserved.

9 The position of sponges in the animal kingdom

Hadzi (1963) when detailing his views on the origin of the Metazoa commented that in discussion of the origin of sponges in relation to that of other Metazoa it was essential to use 'scientific imagination' or, as it is often termed, 'speculation'. It would be out of place in a treatment such as this, where the stress has been on a modern interpretation of the biology of sponges and on relationships within the Porifera, to dwell in conclusion on 'speculation'. Detailed treatments of metazoan origins and of relationships within the lower Metazoa are given by Hadzi (1963), Jägersten (1972) and Hyman (1940, 1959). Specific discussion of the position of sponges in relation to the lower Metazoa, based on detailed personal knowledge of the phylum, is to be found in the writings of Tuzet (1949, 1963, 1973*b*) and Brien (1967, 1972, 1973). We need only deal briefly with the main questions around which speculation revolves in order to put the issues clearly.

All of the above authors concur in placing the origins of the Porifera within the flagellate Protozoa. This is, however, no agreement on whether the sponge ancestor was of the same or different stock to that which gave rise to the Coelenterata (Cnidaria) and presumably to all other Metazoa. This is a crucial difference of opinion, since if one adopts the view that sponge and coelenterate stocks were divergent at this level, then it is impossible to view the Porifera as a phylum within the Metazoa. Probably the most tenable view is that of Tuzet (1973*b*), who suggests that, while the Coelenterata and Porifera lie on the same line of descent, their point of common origin preceded the differentiation of choanoflagellate stock. She interprets the 'ancestral flagellates' as a group which gave rise to many stocks, among which were those that gave rise separately to the Choanoflagellates, Porifera and Coelenterata. This interpretation provides an explanation of the occurrence of choanocytes of comparable structure in all three groups, and indeed in some other higher organisms (Hemichordata). Only in the stock leading to the Porifera did choanocytes become organized into an internal feeding layer. This is uniquely a poriferan venture in evolution. It also becomes possible using Tuzet's interpretation to view the Porifera as having achieved a metazoan grade of evolution.

This brings us to the major question. Are sponges Metazoa or are the views of Sollas (1884) that they constitute a separate subkingdom, the Parazoa, or Delage (1892) that they are a separate subkingdom, the Enantiozoa, more realistic in the light of modern knowledge?

Sollas based the separation of sponges into the Parazoa fundamentally upon the occurrence of choanocytes in sponges and the similarity of these cells to the individual cells of choanoflagellates. The discovery of *Proterospongia*, a rare, small, planktonic choanoflagellate seemed to provide a model for an intermediate stage in the transition from protozoan to parazoan organization (Hadzi 1963). It has already been noted, however, that choanocytes occur in several metazoan lines and thus their presence in sponges provides inadequate ground on its own for any wide separation of sponges and other multicellular organisms.

The concept of sponges as a subkingdom Enantiozoa has been much more difficult to dispel. Delage was impressed with the fact that, during development from the parenchymella larva in Demospongiae like *Mycale* and *Spongilla*, the flagellated cells which provided for the mobility of the larva underwent inward migration at metamorphosis and finally became the inner feeding layer, the adult choanoderm. The internal cells of the larva were thus left to form the outer lining of the body, in this case the pinacoderm.

Delage placed an emphatic embryological interpretation on this sequence. First, he equated the parenchymella larva with the coelenterate planula. Certainly in overall structure they are alike, both are solid, ciliated marine larvae. However, the planula is formed by a sequence involving a gastrulation by ingression or delamination and the ectoderm and endoderm are established at that time. The external layer of the planula is a true definitive ectoderm. The status of the regions of a sponge parenchymella in terms of germ layers is more difficult to determine in the absence of any recognizable gastrulation process during its formation.

Delage coined the term inversion of layers and stated that sponges were animals which during development reversed the two body layers, thus his use of the Greek term *Enantiozoa*. He claimed that the ectoderm invaginated into the endoderm to form the digestive cavity. This is quite the reverse of the metazoan case.

If indeed this 'inversion' does occur then it seems reasonable to separate the sponges from Metazoa at a very basic level indeed. Several modern authors have adopted this view, notably Hadzi (1963), Codreanu (1970) and Ivanov (1971), although all reserve the right to change the terminology and the interpretations put forward by Delage without making any more recent interpretations of the detailed events which occur during reorganization of sponge larval cells into their adult topography.

Against this view, which effectively maintains that sponges are a completely separate line of multicellular evolution, stands a great deal of evidence that in details of physiology (Reiswig 1971*a*), biochemistry (Junqua, Fayolle and Robert 1975), developmental processes, cellular characteristics, connective tissue structure most notably in the presence of collagen (Garrone, Huc and Junqua 1975) even in the possession of primitive systems of self–non-self recognition (MacLennan 1974), sponges have much in common with other lower metazoans.

Tuzet (1963, 1973*b*) and Brien (1967, 1972) have attempted to establish that the 'inversion of layers' does not occur and to provide an interpretation of the structure of complex sponge larvae which explains the vagaries of cellular migration in spatial terms. They point out that Delage based his theory on a very complex larva. Meewis (1938), however, made a careful study of *Oscarella lobularis*, a very primitive demosponge in which the larva is an amphiblastula. There is no trace in this species of any inversion during development. Tuzet (1963) reiterated her earlier work on the development of *Sycon* and stressed that, even in the stomoblastula stage, potential ectodermal and endodermal cells can be distinguished, The former are posterior in the amphiblastula and non-flagellated, the latter are anterior and flagellated. When the larva settles, it settles on the flagellated (endodermal) pole and the oscule opens in the region of the posterior ectodermal (animal) pole. Tuzet attributes no phylogenetic significance to the curious sequence in which the early blastula turns inside out to place the flagella on the external surface. Earlier workers had likened this process of inversion to that which takes place in the blastula of the Volvovcales and on this basis stressed relationship of sponges to colonial algae.

In such simple larvae, no reversal of potential ectoderm (skin) and endoderm (feeding) layers occurs. Brien (1967) suggested that in complex solid larvae, gastrulation, i.e. the process whereby the definitive position and fate of the body regions become fixed, does not take place until fixation and metamorphosis. He interprets all sponge larvae as basically blastulae. In the advanced Demospongiae the potential ectoderm- and mesohyl-producing cells undergo precocious increase in numbers, and the resultant greater volume fills the central space. The resulting structure is the parenchymella. When fixation of this type of larva occurs it is not physically possible for the external layer, the future choanoderm, to invaginate the solid ectoderm (pinacoderm) and mesohyl region. Thus at metamorphosis the flagellated layer disperses, some cells migrate inward singly or in groups to form the choanoderm and others are shed or phagocytosed. This process is viewed as an ontogenetic response in a highly specialized and differentiated dispersal phase, not as a basic and general reproductive peculiarity of sponges.

In Brien's view this process equates with gastrulation, since it results in the future choanoderm and pinacoderm taking up their definitive relationship to each other. Consequently the sponge larva, in which this rearrangement does not take place until metamorphosis, cannot be equated in developmental terms with a coelenterate planula which is a post-gastrulation dispersal phase. If the parenchymella and planula are not to be equated, then the different positions *vis-à-vis* the pole of the larva occupied by ectoderm and endoderm in these two structures is of no phylogenetic significance. Brien had a compendious knowledge of the details of invertebrate development and his views deserve careful consideration. He was also an expert on sponges, as was Tuzet.

There is another possibility. The flagellated larval cells in highly differentiated parenchymellae could be larval cells only, expended at settlement and never migrating through the ectodermal–mesohyl cell mass. Certainly in *Spongilla* this is the case, choanocyte chambers differentiate from cells of the central mass while the larval epithelium is still in place. In this case no 'reversal' or migration occurs to achieve the adult arrangement. There is merely a differentiation of the adult choanocytes from larval archaeocytes. I believe that this pattern, where the larval flagellated cells are in great part shed or phagocytosed, will prove to be more general in sponges. In such cases there is no critical developmental event which equates with gastrulation. However, perhaps there is no need to go as far as Brien in attempting to make sponge development conform in detail with that of other groups. All authors agree that sponges have had a long independent history; detailed developmental patterns are not comparable within the phylum and could far less be expected to be comparable to those of other lower metazoans.

If 'inversion of layers' is not a basic attribute of sponge development then there is little reason to separate sponges from other metazoan stock. In conclusion we should stress that to Delage 'inversion of layers' meant that the primary germ layers were recognizable in sponges and that they behaved in quite the reverse way at gastrulation to those of other higher invertebrates. More modern workers do not agree that the layers particularly in complex form equate in detail with those of other Metazoa and they interpret the details of the spatial reorganization of cells at metamorphosis as a response to physical factors involved in cell packing and as having only ontogenetic significance. After all, within that homogeneous group the Vertebrata gastrulation patterns differ radically in response to the same factors. I believe this is the better way to view the sponge situation.

Tuzet (1973*b*) summarizes the position which most sponge biologists would probably hold today by saying that sponges, like the Coelentarata, are diploblastic, acoel omate Metazoa. The two groups represent distinct stocks stemming from a common origin which of course is unknown.

General reference works on sponges

Borojevic, R., Fry, W. G., Jones, W. C., Levi, C., Rasmont, R., Sara, M. and Vacelet, J., 1967. Mise au point actuelle de la terminologie des Éponges. *Bull. Mus. Hist. nat. Paris*, 2nd ser., **39,** 1224–35.

Fry, W. G., 1970. *The Biology of the Porifera. Symp. zool. Soc. Lond.*, **25.** Academic Press, London and New York, 512 pp.

Grasse, P.-P., 1973. *Traité de Zoologie. Anatomie, Systèmatique, Biologie.* Tome 3 *Spongiaires*. Masson et Cie, Paris, 716 pp.

Harrison, F. W. and Cowden, R. R., 1976. *Aspects of Sponge Biology*. Academic Press, New York, San Francisco, London, 354 pp.

Hyman, L. H., 1940. *The Invertebrates.* Vol. 1. *Protozoa through Ctenophora.* McGraw-Hill, New York, 726 pp.

Hyman, L. H., 1959. *The Invertebrates.* Vol. 5. McGraw-Hill, New York, 783 pp.

Minchin, E. A., 1900. Sponges. In *A Treatise in Zoology*. Pt 2. Ed. E. R. Lankester, Adam and Charles Black, London, pp. 1–178.

Bibliography

Ackermann, D. and List, P. H., 1959. Über das Vorkommen von Taurobetain, Taurin und Inosit in Riesenkieselschwamm. *Hoppe-Seyler's Z. physiol. Chem.*, **317,** 78–81.

Agrell, I., 1951. Enzymes and cell differentiation in sponges, *Ark. Zool.*, **3,** 325–31.

Baer, F. M. and Owre, H. B., 1968. *The Free-living Lower Invertebrates.* Macmillan, New York, 229 pp.

Baker, J. T. and Murphy-Steinmann, V., 1976. *Compounds from Marine Organisms.* In *Handbooks of Marine Science,* sect. 2, vol. 1. CRC Press, Cleveland, Ohio, 400 pp.

Bergmann, W., 1949. Comparative biochemical studies on the lipids of marine invertebrates, with special reference to the sterols. *J. Mar. Res.*, **8,** 137–76.

Bergmann, W., 1962. Sterols: their structure and distribution. *Comprehensive Biochemistry.* Vol. 3. Ed. M. Florkin and H. S. Mason, Academic Press, New York and London, pp. 103–62.

Bergmann, W. and Feeney, R. J., 1950. The isolation of a new thymine pentoside from sponges. *J. Am. chem. Soc.*, **72,** 2805.

Bergmann, W., Watkins, J. C. and Stempien, M. F., 1957. Contributions to the study of marine products. 45. Sponge nucleic acids. *J. org. Chem.*, **22,** 1308–13.

Bergquist, P. R., 1968. The marine fauna of New Zealand. Porifera, Demospongiae. Part 1 (Tetractinomorpha and Lithistida). *NZ Dep. Sci. ind. Res. Bull.*, **188,** 1–106.

Bergquist, P. R., 1972. Phylum Porifera. In *Textbook of Zoology.* Invertebrates. Ed. A. J. Marshall and W. D. Williams, Macmillan, New York, 874 pp.

Bergquist, P. R. and Bedford J. J. (in press). The incidence of antibacterial activity in marine Demospongiae, systematic and geographic considerations. *Mar. Biol.*

Bergquist, P. R. and Hartman, W. D., 1969. Free amino acid patterns and the classification of the Demospongiae. *Mar. Biol.*, **3,** 247–68.

Bergquist, P. R. and Hofheinz, W. (in press). Sterol composition and the classification of the Demospongiae. *Comp. Biochem. Physiol.*

Bergquist, P. R. and Hogg, J. J., 1969. Free amino acid patterns in Demospongiae: a biochemical approach to sponge classification. *Cah. Biol. mar.*, **10,** 205–20.

Bergquist, P. R. and Sinclair, M. E., 1968. The morphology and behaviour of larvae of some intertidal sponges. *NZ J. mar. freshwater Res.*, **2,** 426–37.

Bergquist, P. R. and Sinclair, M. E., 1973. Seasonal variation in settlement and spiculation of sponge larvae. *Mar. Biol.*, **20,** 35–44.

Bergquist, P. R., Sinclair, M. E. and Hogg, J. J., 1970. Adaptation to intertidal existence: reproductive cycles and larval behaviour in Demospongiae. *Symp. zool. Soc. Lond.*, **25,** 247–71.

Bertrand, J.-C. and Vacelet, J., 1971. L'association entre éponges cornées et bactéries. *C.r. hebd. Séanc. Acad. Sci. Paris*, **273,** 638–41.

Bidder, G. P., 1898. The skeleton and classification of calcareous sponges. *Proc. R. Soc.*, **64,** 61–76.

Bidder, G. P., 1923. The relation of the form of a sponge to its currents. *Q. Jl microsc. Sci.*, **67,** 293–323.

Borojevic, R., 1966*a*. Éponges calcaires des côtes de France. II. Le genre *Ascandra* Haeckel emend. *Arch. Zool. exp. gén.*, **107,** 357–68.

Borojevic, R., 1966*b*. Étude expérimentale de la différentiation des cellules de l'éponge au cours de son développement. *Devl Biol.*, **14,** 130–53.

Borojevic, R., 1967. La ponte et le développement de *Polymastia robusta* (Demosponges). *Cah. Biol. mar.*, **7,** 1–6.

Borojevic, R., 1968. Éponges calcaires des côtes de France. IV. Le genre *Ascaltis* Haeckel emend. *Arch. Zool. exp. gén.*, **109,** 193–210.

Borojevic, R., 1969. Étude de développement et de la différentiation cellulaire d'éponges calcaires calcinéennes (genres *Clathrina* et *Ascandra*). *Ann. Embryol. morphog.*, **2,** 15–36.

Borojevic, R., 1970*a*. Différentiation cellulaire dans l'embryogenèse et la morphogenèse chez les spongiaires. *Symp. zool. Soc. Lond.*, **25,** 467–90.

Borojevic, R., 1970*b*. Éponges calcaires des côtes de France. III. Discussion sur la taxonomie des éponges calcaires: *Aphroceras ensata* (Bowerbank) et *Ute gladiata* sp.n. *Arch. Zool. exp. gén.*, **107,** 703–24.

Brien, P., 1967. Les Éponges – leur nature metazoaire – leur gastrulation – leur état colonial. *Ann. Soc. r. zool. Belg.*, **97,** 197–235.

Brien, P., 1969. Les Potamolepides africains. Polyphyletisme des éponges d'eau douce. *Arch. Zool. exp. gén.*, **110,** 527–62.

Brien, P., 1972. Les feuillets embryonnaires des Éponges. *Bull. Acad. r. Belg.*, Ser. 5, **58,** 715–32.

Brien, P., 1973. Les Demosponges. Morphologie et reproduction. In *Traité de Zoologie*. III. *Spongiaires*. Ed. P.-P. Grasse, Masson et Cie, pp. 133–461.

Brien, P. and Meewis, H., 1938. Contribution à l'étude de l'embryogenèse des Spongillides. *Arch. Biol. Paris*, **49,** 177–250.

Brøndsted, H. V., 1936. Entwicklungsphysiologische Studien über *Spongilla lacustris* L. *Acta zool.*, **17,** 75–172.

Burton, M., 1931. The interpretation of the embryonic and post-larval characters of certain tetraxonid sponges with observations on post-larval growth stages in some species. *Proc. zool. Soc. Lond.*, 511–25.

Buss, L. W., 1976. Better living through chemistry: the relationship between allelochemical interactions and competitive networks. In *Aspects of Sponge Biology*. Ed. F. W. Harrison and R. R. Cowden, Academic Press, New York, pp. 315–27.

Chen, W.-T., 1976. Reproduction and speciation in *Halisarca*. In *Aspects of Sponge Biology*. Ed. F. W. Harrison and R. R. Cowden, Academic Press, New York, London, pp. 113–39.

Cimino, G., de Stefano, S. and Minale, L., 1972. Prenylated quinones in marine sponges: *Ircinia* sp. *Experientia*, **38,** 1401, 1402.

Cimino, G., de Stefano, S., Minale, L. and Sodano, G., 1975. Metabolism in Porifera. III. Chemical patterns and the classification of the Demospongiae. *Comp. Biochem. Physiol.*, **50***B*, 279–85.

Codreanu, R., 1970. Grands problèmes controverses de l'évolution phylogénétique des Metazoaines. *Annls Biol.*, **9,** 671–709.

Cohen, S. S., 1963. Sponges, cancer chemotherapy, and cellular aging. *Perspect. biol. Med.*, **6,** 215–27.

Connes, R., 1975. Mode de formation de certains systèmes membranaires au niveau des plaquettes, vitellines des thesocytes d'une Demosponge marine: *Suberites domuncula* (Olivi) Nardo. *C.r. hebd. Acad. Séanc. Sci. Paris*, **281,** 1851–4.

Connes, R., Diaz, J.-P. and Paris, J., 1971. Choanocytes et cellule centrale chez la Demosponge *Suberites massa* Nardo. *C.r. hebd. Séanc. Acad. Sci. Paris*, **273,** 1590–3.

Curtis, A. S. G., 1970. Problems and some solutions in the study of cellular aggregation. *Symp. zool. Soc. Lond.*, **25,** 335–52.

Curtis, A. S. G. and Van de Vyver, G., 1971. The control of cell adhesion in a morphogenetic system. *J. Embryol. exp. Morphol.*, **26,** 295–312.

Dayton, P. K., Robilliard, G. A. and Paine, R. T., 1970. Benthic faunal zonation as a result of Anchor Ice at McMurdo Sound, Antarctica. *Antarct. Ecol.* Ed. M. W. Holdgate, **1,** 244–58.

Dayton, P. K., Robilliard, G. A., Paine, R. T. and Dayton L. B., 1974. Biological accommodation in the benthic community at McMurdo Sound, Antarctica. *Ecol. Monogr.*, **44,** 105–28.

Delage, Y., 1892. Embryogenèse des éponges siliceuses. *Arch. Zool. exp. gén.*, **10,** 345–498.

De Laubenfels, M. W., 1932. Physiology and morphology of Porifera (*Iotrochota birotulata*) Higgin. *Publ. Carnegie Inst. Wash.*, **435,** 37–66.

De Laubenfels, M. W., 1936. A discussion of the sponge fauna of the Dry Tortugas in particular and the West Indies in general, with material for a revision of the families and orders of the Porifera. *Publ. Carnegie Inst. Wash.*, **467,** *Pap. Tortugas Lab.*, **30,** 1–225.

De Laubenfels, M. W., 1955. Porifera. In *Treatise on Invertebrate Paleontology*. Part E. *Archaeocyatha and Porifera*. Ed. R. C. Moore. Geological Society of America and University of Kansas Press, Lawrence, Kansas, pp. E21–E112.

Dendy, A., 1905. Report on the sponges collected by Professor Herdman, at Ceylon, in 1902. *Ceylon Pearl Oyster Fisheries, Supplementary Reports*, **18,** 1–246.

De Rosa, M., Minale, L. and Sodano, G., 1973. Metabolism in Porifera. II. Distribution of sterols. *Comp. Biochem. Physiol.*, **46*B*,** 823–37.

Devos, C., 1965. Le bourgeonnement externe de l'Éponge *Mycale contarenii* Martens. *Bull. Mus. Hist. nat. Paris*, **37,** 534–55.

De Vos, L., 1971. Étude ultrastructurale de la gemmulogenèse chez *Ephydatia fluviatilis*. I. Le vitellus – formation – teneur en arn et glycogène. *J. Microsc.*, **10,** 283–304.

Diaz, J., Connes, R. and Paris, J., 1973. Origine de la lignée germinale chez une Demosponge de l'étang de Thau: *Suberites massa* Nardo. *C.r. hebd. Séanc. Acad. Sci. Paris*, **277,** 661–4.

Duboscq, O. and Tuzet, O., 1937. L'ovogenèse, la fecondation et les premiers stades du développement des Éponges calcaires. *Arch. Zool. exp. gén.*, **79,** 157–316.

Efremova, S. M., 1967. The cell behaviour of the freshwater sponge *Ephydatia fluviatilis*. A time-lapse microcinematography study. *Acta biol. hung.*, **18,** 37–46.

Elvin, D., 1971. Growth rates of the siliceous spicules of the fresh-water sponge *Ephydatia mulleri* (Lieberkühn). *Trans. Am. microsc. Soc.*, **90,** 219–24.

Endo, Y., Watanabe, Y. and Tamura-Hiramoto, S., 1970. Fertilization and development of *Tetilla serica*, a tetraxonian sponge. *Jap. J. exp. Morphol.*, **21,** 40–60.

Evans, R., 1901. A description of *Ephydatia blembingia*, with an account of the formation and structure of the gemmule. *Q. Jl microsc. Sci.*, **44,** 71–110.

Feige, W., 1969. Die Feinstruktur der Ephithelien von *Ephydatia fluviatilis*. *Zool. Jl* (*Anat.*), **86,** 177–237.

Fell, P. E., 1969. The involvement of nurse cells in oogenesis and embryonic development in the marine sponge *Haliclona ecbasis*. *J. Morphol.*, **127,** 133–49.

Fell, P. E., 1970. The natural history of *Haliclona ecbasis* De Laubenfels, a siliceous sponge of California. *Pacific Sci.*, **24,** 381–6.

Fell, P. E., 1974. Porifera. In *Reproduction of Marine Invertebrates*, Vol 1. Ed. A. C. Giese and J. S. Pearse, Academic Press, New York, pp. 51–132.

Finks, R. M., 1967. The structure of *Saccospongia laxata* Bassler (Ordovician) and the phylogeny of the Demospongiae. *J. Palaeontol.*, **41,** 1137–49.

Finks, R. M., 1970. The evolution and ecologic history of sponges during Palaeozoic times. *Symp. zool. Soc. Lond.*, **25,** 3–22.

Finks, R. M. and Hill, D., 1967. Porifera and Archaeocyatha. In *The Fossil Record. A Symposium with Documentation*. Geological Society of London, London, pp. 333–45.

Fry, W. G., 1971. The biology of larvae of *Ophlitaspongia seriata* from two North Wales populations. Fourth *European Marine Biology Symposium.*

Ed. D. J. Crisp, Cambridge University Press, Cambridge, pp. 155–78.

Garrone, R., 1971. Fibrogenèse chez l'Éponge *Chondrosia reniformis* Nardo (Demosponge Tetractinellide). Ultrastructure et function des lophocytes. *C. r. hebd. Séanc. Paris,* **372,** 1832–5.

Garrone, R., 1974. Ultrastructure d'une 'gemmule armée' planctonique d'éponge Clionidae. *Arch. Anat. microsc. Morph. exp.,* **63,** 163–82.

Garrone, R., Huc, A. and Junqua, S., 1975. Fine structure and physiochemical studies on the collagen of the marine sponge *Chondrosia reniformis* Nardo. *J. Ultrastruct. Res.,* **52,** 261–75.

Garrone, R. and Pottu, J., 1973. Collagen biosynthesis in sponges: elaboration of spongin by spongocytes. *J. submicrosc. Cytol.,* **5,** 199–218.

Gilbert, J. J. and Simpson, T. L., 1976. Sex reversal in a freshwater sponge. *J. exp. Zool.,* **195,** 145–51.

Goreau, T. F. and Hartman, W. D., 1963. Boring sponges as controlling factors in the formation and maintenance of coral reefs. *American Association for the Advancement of Science Publ.* no. **75,** 25–54.

Gressinger, J. M., 1971. Étude des Renierides de Méditerranée (Demosponges Haplosclerides). *Bull. Mus. Hist. nat. Paris,* **3,** 98–180.

Hadzi, J., 1963. *The Evolution of the Metazoa.* Pergamon Press, Oxford, London, New York, Paris, 499 pp.

Hadzi, J., 1966. Vprasanje individualitete pri spuzvah (Le problème de l'-individualité chez les éponges). *Slov. Akad. Znan Umet.,* **9,** 167–204.

Haeckel, E., 1872. *Die Kalkschwämme.* 3 vols. Georg Reimer, Berlin.

Hammen, C. S. and Florkin, M., 1968. Chemical composition and intermediary metabolism – Porifera. In *Chemical Zoology.* Vol. 2. Ed. M. Florkin and B. T. Scheer, Academic Press, New York, pp. 53–64.

Harrison, F. W., 1972. The nature and role of the basal pinacoderm of *Corvomeyenia carolenensis* Harrison. (Porifera: Spongellidae). A histochemical and developmental study. *Hydrobiology,* **39,** 495–508.

Hartman, W. D., 1958*a.* Natural history of the marine sponges of southern New England. *Peabody Mus. Nat. Hist., Yale Univ. Bull.,* **12,** 155 pp.

Hartman, W. D., 1958*b.* A re-examination of Bidder's classification of the Calcarea. *Syst. Zool.,* **7,** 97–110.

Hartman, W. D., 1969. New genera and species of coralline sponges (Porifera) from Jamaica. *Postilla,* **137,** 1–39.

Hartman, W. D. and Goreau, T. F., 1970. Jamaican coralline sponges: their morphology, ecology and fossil relatives. *Symp. zool. Soc. Lond.,* **25,** 205–43.

Hartman, W. D. and Goreau, T. F., 1975. A Pacific tabulate sponge, living representative of a new order of scleroponges. *Postilla,* **167,** 1–13.

Hartman, W. D. and Reiswig, H. M., 1973. The individuality of sponges. In *Animal Colonies.* Ed. R. S. Boardman, A. H. Cheetham and W. A. Oliver, Dowden, Hutchinson & Ross Inc., Stroudsburg, Pa., pp. 567–84.

Hazelhoff, E. H., 1938. Über die Ausnutzung des Sauerstoffs bei verschiedenen Wassertieren. *Z. vergl. Physiol.,* **26,** 306–27.

Henkart, P., Humphreys, S. and Humphreys, T., 1973. Characterization of sponge aggregation factor. A unique proteoglycan complex. *Biochemistry*, **12,** 3045–50.

Hentschel, E., 1923. Porifera. In *Handbuch der Zoologie*. Vol. 5. Ed. W. Kükenthal, Berlin, pp. 307–418.

Humphreys, T., 1963. Chemical dissolution and *in vitro* reconstruction of sponge cell adhesions. I. Isolation and functional demonstration of components involved. *Devl Biol.*, **8,** 27–47.

Humphreys, T., 1970. Species specific aggregation of dissociated sponge cells. *Nature, Lond.*, **228,** 685–6.

Hutchinson, G. E., 1970. Living fossils. *Am. Scient.*, **58,** 531–5.

Hyman, L. H., 1925. Respiratory differences along the axis of the sponge *Grantia*. *Biol. Bull.*, **48,** 379–88.

Ijima, I., 1901. Studies on the Hexactinellida, contrib. I. Euplectellidae. *J. Coll. Sci. Imp. Uni. Tokyo*, **15,** 1–299.

Ivanov, A. V., 1971. Embryology of sponges (Porifera) and their position in the animal kingdom. (In Russian.) *J. Biol. Gen. URSS*, **32,** 557–72.

Jackson, J. B. C. and Buss, L., 1975. Allelopathy and spatial competition among coral reef invertebrates. *Proc. natn. Acad. Sci. USA*, **72,** 5160–3.

Jacobsen, J. G. and Smith, L. H., 1968. Biochemistry and physiology of taurine and taurine derivates. *Physiol. Rev.*, **48,** 424–511.

Jägersten, G., 1972. *Evolution of the Metazoan Life Cycle. A Comprehensive Theory*. Academic Press, London, New York, 282 pp.

John, H. A., Campo, M. S., Mackenzie, A. M. and Kemp, R. B., 1971. Role of different sponge cell types in species specific cell aggregation. *Nature New Biol.*, **230,** 126–8.

Jones, W. C., 1962. Is there a nervous system in sponges? *Biol. Rev.*, **37,** 1–50.

Jones, W. C., 1966. The structure of the porocytes in the calcareous sponge *Leucosolenia complicata* (Montagu). *Jl R. microsc. Soc.*, **85,** 53–62.

Jones, W. C., 1970. The composition, development, form and orientation of calcareous sponge spicules. *Symp. zool. Soc. Lond.*, **25,** 91–123.

Jones, W. C. and Jenkins, D. A., 1970. Calcareous sponge spicules: a study of magnesium calcites. *Calcif. Tissue Res.*, **4,** 314–29.

Jørgensen, C. B., 1955. Quantitative aspects of filter feeding in invertebrates. *Biol. Rev.*, **30,** 391–454.

Jørgensen, C. B., 1966. *The Biology of Suspension Feeders*. Pergamon Press, Oxford, 357 pp.

Junqua, S., Fayolle, J. and Robert, L., 1975. Structural glycoproteins from sponge intercellular matrix. *Comp. Biochem. Physiol.*, **50*B*,** 305–9.

Kilian, E. F., 1952. Wasserstromung und Nahrungsaufnahene beim Susswasserschwamm. *Ephydatia fluviatilis*. *Z. vergl. Physiol.*, **34,** 407–47.

Kilian, E. F., 1964. Zur biologie der einheimischen Spongilliden ergebnisse und probleme. Unter besonderer Berucksichtigung eigener Untersuchungen. *Zool. Beitr.*, **10,** 85–159.

Kirkpatrick, R., 1910. On a remarkable pharetronid sponge from Christmas Island. *Proc. R. Soc.*, **83,** 124–33.

Kirkpatrick, R., 1911. On *Merlia normani*, a sponge with a siliceous and calcareous skeleton. *Q. Jl microsc. Sci.*, **56,** 657–702.

Kittredge, J. S., Simonsen, D. G., Roberts, E. and Jelinek, B., 1962. Free amino acids of marine invertebrates. In *Amino Acid Pools.* Ed. J. T. Holden, Elsevier Amsterdam, pp. 176–86.

Kittredge, J. S., Takahashi, F. T., Lindsey, J. and Lasker, R., 1974. Chemical signals in the sea: marine allelochemics and evolution. *Fishery Bull.*, **72,** 1–11.

Krukenberg, C. F. W., 1882. Die Pigmente, ihre Eigenschaften, ihre Genese und ihre Metamorphosen bei den wirbellosen Tieren. In *Vergleichend-physiologische Studien.* R 2, Abt 3. Winter, Heidelberg, pp. 1–115.

Ledger, P. W., 1975. Septate junctions in the calcareous sponge *Sycon ciliatum. Tissue and Cell*, **7,** 13–18.

Lendenfeld, R. von, 1889. *A Monograph of the Horny Sponges.* Trübner & Co., London, 936 pp.

Lentz, T. L., 1966. Histochemical localization of neurohumors in a sponge. *J. exp. Zool.*, **162,** 171–80.

Leveaux, M., 1939. La formation des gemmules chez les Spongillidae. *Ann. Soc. r. zool. Belg.*, **70,** 53–96.

Leveaux, M., 1941. Contribution à l'étude histologique de l'ovogenèse et de la spermatogenèse des Spongillides. *Ann. Soc. r. zool. Belg.*, **72,** 126–49.

Levi, C., 1956. Étude des *Halisarca* de Rosocoff. Embryologie et systématiques des Demosponges. *Arch. Zool. exp. gén.*, **93,** 1–181.

Levi, C., 1957. Ontogeny and systematics in sponges. *Syst. zool.*, **6,** 174–83.

Levi, C., 1960. Les Demosponges des côtes de France. 1. Les Clathriidae. *Cah. Biol. mar.*, **1,** 47–87.

Levi, C., 1967. Les fibres segmentées intracellulaires d'*Haliclona elegans* Bow (Demosponge Haplosclerida). *Arch. Zool. exp. gén.*, **108,** 611–16.

Levi, C., 1973. Systématique de la classe des Demospongiaria (Demosponges). In *Traité de Zoologie.* III. *Spongiaires.* Ed. P.-P. Grasse, Masson et Cie, pp. 37–631.

Litchfield, C., 1976. What is the significance of coloration in sponges? In *Aspects of Sponge Biology.* Ed. F. W. Harrison and R. P. Cowden, Academic Press, New York, pp. 28–32.

Maas, O., 1893. Die Embryonal-Entwicklung und metamorphose der Cornacuspongien. *Zool. Jb.*, **7,** 331–448.

MacLennan, A. P., 1974. The chemical bases of taxon-specific cellular reaggregation and 'self'–'not-self' recognition in sponges. *Archs Biol.*, **85,** 53–90.

MacMunn, C. A., 1900. On spongioporphyrin: the pigment of *Suberites wilsoni. Q. Jl microsc. Sci.*, **43,** 337–49.

Meewis, H., 1938. Contribution à l'étude de l'embryogenèse des Myxospongidae *Halisarca lobularis* (Schmidt). *Archs Biol.*, **50,** 3–66.

Meewis, H., 1948. Contribution à l'étude histologique des spongiaires. *Ann. Soc. r. zool. Belg.*, **79,** 5–36.

Moret, L., 1952. Embranchement des spongiaires (Porifera, Spongiata). In *Traité de Paleontologie* I. Ed. J. Piveteau, Masson et Cie, Paris, pp. 334–74.

Muller, W. E. G. and Zahn, R. K., 1973. Purification and characterization of a species-specific aggregation factor in sponges. *Expl. Cell Res.*, **80,** 95–104.

Neumann, A. C., 1966. Observations on coastal erosion in Bermuda and measurements of the boring rate of the sponge, *Cliona lampa. Limnol. Oceanogr.*, **11,** 92–108.

Nicol, J. A. C., 1960. *The Biology of Marine Animals.* Pitman, Bath, 707 pp.

Okada, Y., 1928. On the development of a hexactinellid sponge *Farrea sollasii. Tokyo Univ. Fac. Sci. J.* Sect. IV, **2,** 1–27.

Pavans de Ceccatty, M., 1958. La mélanization chez quelques éponges calcaires et siliceuses: ses rapports avec le système reticulo-histocytaire. *Arch. Zool. exp. gén.*, **96,** 1–51.

Pavans de Ceccatty, M., 1974*a*. Coordination in sponges. The foundations of integration. *Am. Zool.*, **14,** 895–903.

Pavans de Ceccatty, M., 1974*b*. The origin of the integrative systems: a change in view derived from research on coelenterates and sponges. *Perspect. biol. Med.*, **17,** 379–90.

Pavans de Ceccatty, M., Thiney, Y. and Garrone, R., 1970. Les bases ultra-structurales des communications intercellulaires dans les oscules de quelques éponges. *Symp. zool. Soc. Lond.*, **25,** 449–66.

Polejaeff, N., 1883. Report on the Calcarea dredged by H.M.S. *Challenger* during the years 1873–76. *Rep. Sci. Results Voyage Challenger Zool.*, **8,** 1–76.

Pourbaix, N., 1933. Recherches sur la nutrition des Spongiaires. *Inst. Esp. Oceanogr. Notas y Resumenes* Ser. II, **69,** 2–42.

Putter, A., 1914. Der stoffwechsel der Kieselschwämme. *Z. allg. Physiol.*, **16,** 65–114.

Rasmont, R., 1956. La gemmulation des Spongillides. (IV) Morphologie de la gemmulation chez *Ephydatia fluviatilis* et *Spongilla fragilis. Annls Soc. r. zool. Belg.*, **86,** 349–87.

Rasmont, R., 1959. L'ultrastructure des choanocytes d'éponges. *Annls Sci. nat.* (*Zool.*), 12e Ser., 253–62.

Rasmont, R., 1962. The physiology of gemmulation in freshwater sponges. *Symp. Soc. Study Devl Growth*, **30,** 3–25.

Rasmont, R., 1963. Le rôle de la taille et de la nutrition dans le déterminisme de la gemmulation chez les Spongillides. *Devl Biol.*, **8,** 243–71.

Rasmont, R., 1965. Existence d'une regulation biochimique de l'eclosion des gemmules chez les Spongillides. *C.r. hebd. Séanc. Acad. Sci. Paris*, **261,** 845–6.

Rasmont, R., 1970. Some new aspects of the physiology of freshwater sponges. *Symp. zool. Soc. Lond.*, **25,** 415–22.

Reiswig, H. M., 1970. Porifera: sudden sperm release by tropical Demospongiae. *Science*, **170,** 538–9.

Reiswig, H. M., 1971*a*. *In situ* pumping activities of tropical Demospongiae. *Mar. Biol.*, **9,** 38–50.

Reiswig, H. M., 1971*b*. Particle feeding in natural populations of three marine Demosponges. *Biol. Bull.*, **141,** 568–91.

Reiswig, H. M., 1973. Population dynamics of three Jamaican Demospongiae. *Bull. mar. Sci.*, **23,** 191–226.

Reiswig, H. M., 1974. Water transport, respiration and energetics of three tropical marine sponges. *J. exp. mar. Biol. Ecol.*, **14,** 231–49.

Reiswig, H. M., 1976. Natural gamete release and oviparity in Caribbean Demospongiae. In *Aspects of Sponge Biology*. Ed. F. W. Harrison and R. R. Cowden. Academic Press, New York, San Francisco, London, pp. 99–112.

Rezvoi, P. D., Zhuravleva, I. T. and Koltun, V. M., 1971. Phylum Porifera. In *Fundamentals of Paleontology*. Vol. 1, Pt II. Porifera, Archaeocyatha, Coelenterata, Vermes. Ed. Yu. A. Orlov and B. S. Sokolov, Israel Program for Scientific Translations Ltd, Jerusalem, pp. 5–97.

Ridley, S. O. and Dendy, A., 1887. Report on the Monaxonida collected by H.M.S. *Challenger* during the years 1873–1876. *Rep. Sci. Results Voyage Challenger Zool.*, **20,** 1–275.

Rigby, J. K., 1976. Some observations on occurrence of Cambrian Porifera in Western North America and their evolution. *Geol. Stud.*, **23,** 51–60.

Rigby, J. K. and Nitecki, M. H., 1975. An unusually well preserved heteractinid sponge from the Pennsylvanian of Illinois and a possible classification and evolutionary scheme for the Heteractinida. *J. Paleontol.*, **49,** 329–39.

Roche, J. and Robin, Y., 1954. Sur les phosphagènes des éponges. *C.r. Séanc. Soc. Biol.*, **148,** 1541–3.

Rosenfeld, F., 1970. Inhibition du développement des gemmules de Spongillides: spécificité et moment d'action de la gemmulostasine. *Archs Biol.*, **81,** 193–214.

Rosenfeld, F., 1971. Éfféts de la perforation de la coque des gemmules d'*Ephydatia fluviatilis* (Spongillides) sur leur développement ulterieur en presence de gemmulostasine. *Archs Biol.*, **82,** 102–13.

Rützler, K., 1965. Substratstabilität im marinen Benthos als ökologischer Faktor, dargestellt am Beispiel adriatischer Porifera. *Int. Rev. ges. Hydrobiol.*, **50,** 281–92.

Rützler, K., 1970. Spatial competition among Porifera: solution by epizoism. *Oecologia*, **5,** 85–95.

Rützler, K., 1975. The role of burrowing sponges in bioerosion. *Oecologia*, **19,** 203–16.

Rützler, K. and Rieger, G., 1973. Sponge burrowing: fine structure of *Cliona lampa* penetrating calcareous substrate. *Mar. Biol.*, **21,** 144–62.

Sara, M., 1970. Competition and cooperation in sponge populations. *Symp. zool. Soc. Lond.*, **25,** 273–84.

Sara, M., 1971. Ultrastructural aspects of the symbiosis between two species of the genus *Aphanocapsa* (Cyanophyceae) and *Ircinia variabilis* (Demospongiae). *Mar. Biol.*, **11,** 214–21.

Sara, M. and Vacelet, J., 1973. Écologie des Demosponges. In *Traité de Zoologie*. Vol. 3. *Spongiaires*. Ed. P.-P. Grasse. Masson et Cie, Paris, pp. 462–576.

Schmidt, I., 1970. Phagocytose et pinocytose chez les Spongillidae. *Z. Vergl. Physiol.*, **66,** 398–420.

Schmidt, O., 1864. *Supplement der Spongien des adriatischen Meeres, enthattend die Histologie und systematische Erganzungun.* W. Engelmann, Leipzig, 48 pp.

Schmidt, O., 1870. *Grundzüge einer Spüongien-fauna des Atlantischen Gebietes.* Leipzig, 88 pp.

Schulze, F. E., 1878. Untersuchungen über den Bau und die Entwicklung der Spongien. V. Die Metamorphose von *Sycandra raphanus. Z. wiss. Zool.*, **31,** 262–95.

Schulze, F. E., 1880. On the structure and arrangement of the soft parts in *Euplectella aspergillum. Trans. R. Soc. Edinb.*, **29,** 661–73.

Schulze, F. E., 1887. Report on the Hexactinellida collected by H.M.S. *Challenger* during the years 1873–76. *Rep. Sci. Results Voyage Challenger Zool.*, **21,** 1–513.

Simpson, T. L., 1968. The biology of the marine sponge *Microciona prolifera* (Ellis and Solander). II. Temperature-related, annual changes in functional and reproductive elements with a description of larval metamorphosis. *J. exp. mar. Biol. Ecol.*, **2,** 252–77.

Siribelli, L., 1962. Differenze nel ciclo sessuale di popolazioni conviventi di *Axinella damicornis* (Esper) ed *Axinella verrucosa* o.s. (Demospongiae). *Ann. Inst. Mus. Zool. Univ. Napoli*, **14,** 1–8.

Sollas, W. J., 1884. On the development of *Halisarca lobularis* (O. Schmidt). *Q. Jl*, **24,** 603–21.

Sollas, W. J., 1888. Report on the Tetractinellida collected by H.M.S. *Challenger* during the years 1873–1876. *Rep. Sci. Results Voyage Challenger Zool.*, **25,** 1–458.

Stearn, C. W., 1972. The relationship of the stromatoporoids to the sclerosponges. *Lethaia*, **5,** 369–88.

Thiney, Y., 1972. Morphologie and cytochimie ultrastructurale de l'oscule d'*Hippospongia communis* LMK et de sa regeneration. Thèse, Université Claude Bernard, 63 pp.

Topsent, E., 1928. Spongiaires de l'Atlantique et de la Méditerranée provenant des croisières de Prince Albert 1er de Monaco. *Reultats des Campagnes Scientifiques du Prince de Monaco*, **74,** 1–376.

Travis, D. F., François, C. J., Bonar, L. C. and Glimcher, M. J., 1967. Comparative studies of the organic matrices of invertebrate mineralized tissues. *J. Ultrastruct. Res.*, **18,** 519–50.

Turner, R. S. and Burger, M. M., 1973. Involvement of a carbohydrate group in the active site for surface guided reassociation of animal cells. *Nature, Lond.*, **244,** 509–10.

Tuzet, O., 1947. L'ovogenèse et la fecondation de l'éponge calcaire *Leucosolenia* (*Clathrina*) *coriacea* Mont. et de l'éponge siliceuse *Reniera elegans* Bow. *Arch. Zool. exp. gén.*, **85,** 127–48.

Tuzet, O., 1948. Les premiers stades du développement de *Leucosolenia botry-*

oides Ellis et Solander et de *Clathrina* (*Leucosolenia*) *coriacea* Mont. *Annls Sci. nat.* (*Zool.*), **11,** 103–13.

Tuzet, O., 1949. La place des Spongiaires dans la classification. *8th Congr. int. Zool. Paris*, 429–32.

Tuzet, O., 1963. The phylogeny of sponges according to embryological, histological and serological data, and their affinities with the Protozoa and Cnidaria In *The Lower Metazoa. Comparative Biology and Phylogeny*. Ed. E. C. Dougherty, Z. N. Brown, E. D. Hanson and W. D. Hartman, University of California Press, San Francisco, pp. 129–48.

Tuzet, O., 1973*a*. Éponges calcaires. In *Traité de Zoologie*. Vol. 3. *Spongiaires*. Ed. P.-P. Grasse, Masson et Cie, pp. 27–132.

Tuzet, O., 1973*b*. Introduction et place des spongiaires dans la classification. In *Traité de Zoologie*. Vol. 3. *Spongiaires*. Ed. P.-P. Grasse, Masson et Cie, Paris, pp. 1–26.

Tuzet, O., Garrone, R. and Pavans de Ceccatty, M., 1970. Observations ultrastructurales sur la spermatogenèse chez la demosponge *Aplysilla rosea* Schulze (Dendroceratide): une metaplaise exemplaire. *Annls Sci. nat.* (*Zool. et Biol. Anim.*), **12,** 27–50.

Tuzet, O. and Pavans de Ceccatty, M., 1958. La spermatogenèse, l'ovogenèse, la fecondation et les premiers stades du développement d'*Hippospongia communis* LMK (=*H. equina* o.s.). *Bull. Biol. Fr.*, **92,** 1–18.

Tuzet, O., Pavans de Ceccatty, M. and Paris, J., 1963. Les Éponges sont-elles des colonies? *Arch. Zool. exp. gén.*, **102,** 14–19.

Vacelet, J., 1961. The order Pharetronida in Hartman's classification of the Calcarea. *Syst. Zool.*, **10,** 45–7.

Vacelet, J., 1964. Étude monographique de l'Éponge calcaire Pharetronide de Mediterranée *Petrobiona massiliana* Vacelet et Levi. Les Pharetronides actuelles et fossiles. Thèse Docteur des Sciences Naturelles l'Université d'Aix-Marseille, 125 pp.

Vacelet, J., 1967. Descriptions d'Éponges Pharetronides actuelles des tunnels obscurs sous-recfaux de Tulear (Madagascar). *Rec. Trav. St. Mar. End., Suppl.*, **6,** 37–62.

Vacelet, J., 1970. Les éponges Pharetronides actuelles. *Symp. zool. Soc. Lond.*, **25,** 189–204.

Vacelet, J., 1971*a*. L'ultrastructure de la cuticle d'éponges cornées du genre *Verongia*. *J. Microsc.*, **10,** 113–16.

Vacelet, J., 1971*b*. Ultrastructure et formation des fibres de spongine d'éponges cornées *Verongia*. *J. Microsc.*, **10,** 13–32.

Vacelet, J., 1975. Étude en microscopie éléctronique de l'association entre bactéries et spongiaires du genre *Verongia* (Dictyoceratida). *J. Microsc. Biol. cellul.*, **23,** 271–88.

Van de Vyver, G., 1970. La non confluence intraspécifique chez les spongiaires et la notion d'individu. *Annls Embryol. Morphol.*, **3,** 251–62.

Van Weel, P. B., 1949. On the physiology of the tropical fresh water sponge

Spongilla proliferanes. I. Ingestion, digestion and excretion. *Physiol. comp. Oecol.*, **1**, 110–26.

Volkonsky, M., 1930. Les choanocytes des Éponges calcaires. Les phénomènes cytologiques au cours de la digestion intracellulaire. *C.r. Séanc. Soc. Biol.*, **103**, 668–72.

Warburton, F. E., 1966. The behaviour of sponge larvae. *Ecology*, **47**, 672–4.

Watanabe, Y., 1957. Development of *Tethya serica* Lebwohl, a tetraxonian sponge. I. Observations on external changes. *Nat. Sci. Rep. Ochanomizu Univ.*, **8**, 97–104.

Weinbaum, G. and Burger, M. M., 1973. Two component system for surface guided reassociation of animal cells. *Nature, Lond.*, **244**, 510–12.

Weinheimer, A. J. and Spraggins, R. L., 1969. The occurrence of two new prostaglandin derivatives (15-epi-PGA_2 and its acetate methyl ester) in the Gorgonian *Plexaura homomalla*. *Chemistry of Coelenterates*, **15.** *Tetrahedron Lett.*, 5185.

Wells, H. W., Wells, M. J. and Gray, I.E., 1964. Ecology of sponges in Hatteras Harbor, North Carolina. *Ecology*, **45,** 752–67.

Whittaker, R. H. and Feeney P. P., 1971. Allelochemics: chemical interactions between species. *Science*, **171,** 757–70.

Wilson, H. V., 1891. Notes on the development of some sponges. *J. Morphol.*, **5,** 511–19.

Wilson, H. V., 1894. Observations on the gemmule and egg development of marine sponges. *J. Morphol.*, **9,** 277–406.

Yamaguchi, M., 1957. Chemical constitution of renieratene. *Bull. chem. Soc. Japan*, **30,** 979.

Yamaguchi, M., 1958. Chemical constitution of isorenieratene. *Bull. chem. Soc. Japan*, **31,** 51.

Zhuravleva, I. T., 1970. Porifera, Sphinctozoa, Archaeocyathi – their connections. *Symp. zool. Soc. Lond.*, **25,** 41–59.

Ziegler, B. and Rietschel, S., 1970. Phylogenetic relationships of fossil calcisponges. *Symp. zool. Soc. Lond.*, **25,** 23–40.

Glossary

acantho: a prefix meaning spiny, applied to both microscleres and megascleres.

accessory spicule: a category of megasclere supplementary to the primary skeleton. Can be located anywhere in the sponge.

allelochemical: a chemical agent of natural origin involved in interaction between species or individuals.

amphiblastula: a type of larva which possesses a central cavity, and which has two morphologically distinct types of cells, one anterior the other posterior.

apopyle: an exhalant aperture from a choanocyte chamber.

archaeocyte: a totipotent amoeboid cell.

arcuate: a type of chelate microsclere where the teeth are in the form of one to three curved plates.

ascon: a simple sponge with unfolded pinacoderm and choanoderm.

aster: a type of microsclere with several rays originating from the same centre.

astrorhizae: star-like depressions on the surface of the calcareous skeleton of some Sclerospongiae and Stromatoporoidea.

atrium: a cavity into which many exhalant systems empty. A common feature of hollow cylindrical sponges.

axial construction: a type of skeletal organization where some components are condensed to form a dense central region or axis.

axial filament: the protein core around which siliceous spicules are organized.

basal lamina: the attachment surface of a sponge.

carrier cell: a migratory choanocyte which conveys captured sperm to the oocyte.

'cellules en croix': non-flagellated cells disposed in a tetraradial fashion in the amphiblastula of Calcaronea.

central cell: a cell located in the cavity of a choanocyte chamber.

chela: a type of microsclere with a curved axis and recurved teeth at each end.

choanocyte: flagellate cell crowned by a collar of cytoplasmic tentacles. It is responsible for generating the water current.

choanocyte chamber: a cavity enclosed by a group of choanocytes.

choanoderm: the entire surface in any sponge which is lined by choanocytes.

choanosome: the area of the sponge body in which choanocyte chambers are found.

coeloblastula: the simple type of blastula larva found in the Calcinea.

coenosteum: the body of a stromatoporoid.

collencyte: mobile cell responsible for collagen secretion.

conule: an elevation of the surface membrane of a sponge.

cortex: a superficial region of a sponge supported by a special organic or inorganic skeleton.

cribripore: a specialized exhalant structure where several exhalant systems combine to empty into a subsurface cavity.

cystencyte: a polysaccharide-secreting cell with contents enclosed in a single vesicle. Found in fresh-water sponges.

desma: branched, irregular interlocking megasclere found in lithistid sponges.

diamorph: a term sometimes used to designate a cell mass formed as a result of aggregation of dissociated cells. It has a spherical form and a continuous pinacoderm.

echinating spicule: a megasclere which protrudes from a fibre or spicule tract.

ectomesenchyme: a term with wide usage, but when applied to sponges it implies all components of a sponge except for the flagellated cells.

ectosomal spicule: spicules which occur in the ectosomal region of a sponge, sometimes terminating spicule tracts.

ectosome: a superficial region of a sponge which is not supported by any special skeleton.

endosome: all except the ectosomal structures of a sponge.

etching cells: specialized archaeocytes which secrete the chemical agents which enable some sponges to etch calcium carbonate.

extra-axial skeleton: skeletal elements surrounding or arising from an axial region.

factor: a cell surface-active proteoglycan which affects the stability of cell-to-cell adhesion.

founder cells: cells responsible for lengthening rays during secretion of calcareous spicules.

gemmule: a desiccation-resistant asexual reproductive body composed of a mass of archaeocytes charged with reserves and enclosed in a non-cellular protective envelope.

gemmulostasin: inhibitor of gemmule germination.

globoferous cell: cell with a prominent array of paracrystalline components.

gray cell: cell with spherical basiphilic granules and many glycogen rosettes.

isodictyal: a type of skeletal construction where spicules and/or fibres interlock on a regular triangular pattern.

leucon: a sponge in which the flagellated cells are restricted to discrete choanocyte chambers which are dispersed in a thick mesohyl.

lophocytes: extremely mobile collagen-secreting cells which trail attached collagen fibrils.

matrix: the non-cellular ground material of a sponge in which the cellular elements are dispersed.

megasclere: a structural spicule.

mesohyl: the region of a sponge body enclosed between the pinacoderm and the choanoderm.

microgranular cells: cells with cytoplasm charged with small dense granules.

microsclere: a packing or reinforcing spicule, usually of small size, frequently of ornate shape.

monactinal: describes a spicule with one terminal point or actine.

monaxon: a spicule with one axis.

monaxonid: an adjective describing sponges with monaxon spicules.

monocrepid: a monaxonid desma.

myocytes: cells which cause contraction.

nurse cells: cells which migrate toward developing oocytes and when incorporated provide material for further growth.

olynthus: an asconoid, post-settlement stage traversed in the development of calcareous sponges.

oscules: the exhalant apertures of sponges.

ostia: the inhalant openings of a sponge.

parenchymella: a solid, ciliated larva.

periflagellar membrane: a membrane interposed between choanocyte collar tentacles and the apical flagellum.

perispicular spongin: spongin which surrounds spicules.

pinacocytes: cells which line all surfaces except those of the choanocyte chambers. They are differentiated by their position as exo-, endo- or basopinacocytes.

pinocoderm: a layer of pinacocytes.

pith: a central region of more diffuse collagen found within a spongin fibre.

plumoreticulate skeleton: a type of skeletal construction in which fibres or spicule tracts diverge in plumose fashion but still retain cross-connections.

plumose skeleton: a type of skeletal construction in which diverging fibres or spicule tracts show few if any cross-connections.

porocalyx: specialized, sunken, inhalant and exhalant aperture of some Spirophorida.

porocyte: pinacocytes which enclose a pore which functions as an inhalant canal.

primary spicule: the major structural megascleres in any particular sponge.

prosopyle: inhalant opening to a choanocyte chamber.

radial skeleton: a type of skeletal construction where the structural elements diverge from a central point toward the sponge surface.

raphide: a thin, unornamented diactinal microsclere.

reticular membrane: membrane formed by the fused bases of the choanocytes in Hexactinellida.

rhabdiferous cell: cell which secretes mucopolysaccharide.

sclerocyte: cell which secretes spicules.

sieve area: an area in which inhalant apertures are concentrated; often marked by a fringing membrane or a surface depression.

sigma: a type of microsclere in which a single axis is curved or contorted.

silicalemma: a unit membrane which encloses the axial filament of a siliceous spicule.

spherulous cell: cells with multiple large vesicles filled with coarse granular material.

spicule: a discrete element of the sponge skeleton. Usually composed mainly of silica or calcium carbonate but spongin spicules do occur rarely.

spongin: collagenous material deposited in the form of homogeneous fibres or plaques which are often of large size.

spongocoel: the central cavity of a simple asconoid sponge.

spongocyte: cell which secretes spongin.

statoblast: a gemmule of fresh-water sponges belonging to the family Potamolepidae. They lack a complex spongin coat and are covered externally by megascleres.

stauractine: a tetractinal spicule with all four rays in a single plane.

stereogastrula: a general term for solid metazoan larvae; a parenchymella is one type of stereogastrula.

stomoblastula: a period during the development of the calcaronean amphiblastula when the blastula opens and ingests adjacent choanocytes.

sycon: a sponge in which some folding of the choanocyte layer has occurred as well as some superficial thickening of the mesohyl.

tetracladine: an alternative term to tetracrepid; in more frequent use in the paleontological literature.

tetracrepid: a tetraxonid desma.

tetraxonid: a spicule with four axes, e.g. calthrops.

thesocytes: binucleate, highly vitelline archaeocytes of hibernating fresh-water sponge gemmules.

thickener cells: cells responsible for thickening rays during secretion of calcareous spicules.

toxa: a type of microsclere with a single axis flexed in one plane.

triaene: a tetraxonid spicule with three rays shorter than the fourth.

triaxon: a spicule with three axes.

trophocytes: nurse cells involved in the initial stages of gemmule formation in fresh-water sponges.

unguiferate: a type of chelate microsclere in which the teeth are short and discrete, frequently more than three at each end of the shaft.

Index